# ORGANIC MANURE

## Sources, Preparation and Usage in Farming Lands

## The Editor

**Professor S.M. Singh** is an Alumni of University of Lucknow, did Doctorate on "*Structure, development and eco-physiology of two species of earthworms*" in 1989 and taught Zoology from 1990-1993. Since then he is serving this University. He is one of known earthworm biologists with more than thirty one years of illustrious research experience in earthworm ecology, eco-physiology, biodiversity, Vermitechnology and organic farming for sustainable agriculture and twenty five years of PG teaching at the University.

He has been awarded INSA Visiting Scientist and ZSI Fellowships. Prof. Singh has Sixty five research papers in peer reviewed National & International journals to his credit and more than a dozen of popular scientific articles in different Hindi newspapers. He has written 4 books on Earthworms. He has given 102 All India Radio Talks on the environment, waste management, vermicomposting and organic farming. He has completed 4 research projects from various funding agencies (UGC, CSTUP, UP Govt., & NABARD) on Earthworms. He has participated in 41 National Symposia and visited Neuchatel University, Beatenberg (Switzerland) in 2011, Jagiellonian University, Krakow (Poland) in 2006; Cardiff University, Wales (U.K.) in 2002 and Beijing Agricultural University, Beijing (China) in 1995 to participate in the Symposia on Earthworms and discourse on various aspects of Indian earthworms at International forum for the betterment of environmental sustainability.

Prof. Singh organized 6 National Symposia on Earthworm Ecology in the University Campus and in other State and Central Universities of the country & 16 Training & Extension programs at School & Village level for the transfer of Vermitechnology from Lab to Land & at the farmer's door. Prof. Singh has developed a Centre for Vermiculture & Vermicomposting at the University Campus - a research, training and extension unit of the University under Lab. to Land programs and working as Co-ordinator-cum-Incharge of the Center. He is Founder & Secretary of the Society for Earthworm Ecology & Environmental Research (India) and Editor of "VERMECO"- a biannual bulletin on Agricultural Sustainability and Rural Development. At present, Prof. Singh is Dean of the Faculty of Applied Sciences and Dean & Coordinator, Faculty of Medical, Dental and Paramedical Sciences at Mahatma Jyotiba Phule Rohilkhand University, (Uttar Pradesh) Bareilly.

# ORGANIC MANURE
## Sources, Preparation and Usage in Farming Lands


**S.M. Singh**
Dean
Faculty of Applied Sciences
Dean and Coordinator
Faculty of Medical, Dental and Paramedical Sciences
Mahatma Jyotiba Phule Rohilkhand University,
Bareilly-243006, (U.P.), India.



2020
**Associated Publishing Company®**
*A Division of*
**Astral International Pvt. Ltd.**
**New Delhi – 110 002**

ISBN 9789390371112 (Int. Edition)



*Published by* : **Associated Publishing Company®**
*A Division of*
**Astral International Pvt. Ltd.**
– ISO 9001:2015 Certified Company –
4736/23, Ansari Road, Darya Ganj
New Delhi-110 002
Ph. 011-43549197, 23278134
E-mail: info@astralint.com
Website: www.astralint.com

# Preamble

India is one of the leading producers of agricultural and livestock commodities like grains, fruits, vegetables, milk etc., but commercialization of agriculture has had very negative effects on the environment. Extensive dependence on chemical farming has shown its darker side. The entire ecosystem has got deteriorated due to excessive and indiscriminate usage of chemical fertilizers and pesticides. Consumption of chemical fertilizers has been increased >9 times and pesticides by > 2 times since 1971. These high doses have badly affected the physical, chemical and biological parameters of the soil. The farming land has lost its fertility and is demanding larger quantities of fertilizers to be used. Pests are becoming more immune. We have adopted the wrong road of unsustainability in the name of growing more to feed the earth.

Organic farming is a holistic ecologically balanced approach to farming. It is not only about farming without using chemicals but also about the environment, agricultural traditional seeds, animal welfare, soil and water conservation. It is also about you, your family, your health and your welfare. Successful organic farming needs scientific knowledge of ecology and modern biological processes. It reduces toxic load and keeps chemicals out of air, water, soil and form our bodies.

We have to develop the 'Culture' in Agri-'Culture' by minimizing the usage of chemical fertilizers and pesticides and maximizing the usage of organic-manures and organic-pesticides. We could save our "Motherland"- the Earth, and grow healthy, non diseased, toxins-free food grains, fruits, vegetables and fodder by adopting organic farming.

The present scenario to cope with such adverse situation is to produce more and more organic manure by the cheapest and easiest technologies to replace the usages of the chemical fertilizers which are one of the great problems. Generally, it has been observed that our farming communities do not have the idea about the different sources of substrates from which one could easily prepare the organic manure and their usage in the farming system.

The present book is a remedial solution having simple and lucid techniques for the preparation of organic manure from easily available substrates around us where only minimum man power is needed and the money is not required. In this book, six commonly available biodegradable wastes have been taken for the preparation of organic manure. These wastes are flower waste, fruits and vegetable waste, kitchen waste, press mud waste, paper mill sludge waste and fly ash waste. The seventh chapter of the book is concerned with organic agriculture and its need in the today's scenario of farming system. The last chapter is on Bio IPM techniques which could be applied for controlling pests of the crops by treating them with bio-pesticides which are neither harmful to crops and humans nor to the other animals.

**S.M. Singh**
*Department of Animal Science,*
*Mahatma Jyotiba Phule Rohilkhand University,*
*Bareilly-243006,(Uttar Pradesh)India.*

# Contents

# 1

# Flower Waste: A Rich Source for Organic Manure

*A K Sannigrahi*

*Scientist F & Additional Director, Proof & Experimental Establishment, DRDO, Chandipur, Balasore (Odisha)*

## Flower Waste

Thousands of people are visiting everyday to either Temple or Mosque or Church or Gurudwara or any Holy place to pray Almighty to get happiness, peace of mind and success. Hindu devotees generally pay floral tributes in the form of garlands or loose flowers to deities in order to get blessings (Photo-1). These worshipped flowers are thrown away as waste material. These worshipped flowers are a matter of great concern to the Temple Management Authorities since these are either thrown out in to the water bodies (nearby ponds or river) creating severe water pollution or dumped at the roadside garbage-bins causing environmental pollution as well as hampering aesthetic belief of devotees. Flower vendors, big hotels and marriage mandaps also generate a considerable quantity of waste flowers. There are about 1.08 lakh temples in India. Everyday a considerable quantity of floral waste is generated in each temple, quantity depends on number of visitors and also on auspicious days and festivals.

Photo- 1: Flower decorated Shree Siddhivinayak Ganapati

Being organic materials, these can easily be transformed in to good quality manure. Experiment showed that nutrient rich vermicompost can be produced easily within two months from any flower wastes and Temple Management authority can dispose their flower wastes beneficially under employment generation program by converting these flowers to vermicompost which would either be used in Temple garden or be sold to the devotees at a reasonable price. The technique will also help to keep the temple premises clean and healthy. The present article deals with the current scenario of generation of flower waste from famous temples of India, environmental problems on its disposal and different beneficial utilizations of these wastes as important resource material by making agarbatti, by extracting dye or by converting them in to vermicompost and millicompost.

**Photo- 2:** Flowers sold by Flower Vendors

Photo-3: Flower wastes generated in the Farm

Lot of flower vendors (Photo- 2) have their shops near the temples or in market places in every city, who sell flowers not only to devotees but also other people for beautifying marriage pandals and bridegroom's car, for preparing bride's and deity's flower ornaments, for decorating business as well as social parties and events and also for offering as a token of love / respect in the form of stylish bouquets to prominent personalities like beloved guests and elders, achievers, official bosses and political leaders, etc. After completion of functions these flowers are also become waste flowers. Everyday a huge amount of waste is generated at the flower market in the form of damaged or partly dried flowers. During the time of special occasions or pujas, the amount of flower wastes increases leading to major difficulties in terms of clearing them. The Mallick Ghat Flower Market in Howrah (West Bengal) generates everyday about 5 tons of flower wastes composed of rotten flowers, leaves, etc and the monthly cost of clearing those wastes from this flower market is about 1.5 lakh (HT, 2012). During harvesting of flowers, considerable quantity of flowers is also separated out in the field by farmers as non-sellable waste flowers (Photo-3).

## Generation of Waste Flowers in Temples and Other Points

Those flowers offered to deities along with other 'Nirmalayas' are generally thrown away after worshipping as waste materials in open places (Photo- 4), municipality dustbins, temple ponds or in nearby rivers (Photo - 5). Flowers used for decoration or bouquet are also thrown in Municipality dustbin or in dumping ground when the function is over. Flower vendors throw their waste flowers either in dustbin or in nearby drains. Municipal sweepers sometimes burn those flowers wastes inside the dustbin only.

**Photo- 4:** Dumping of flower waste, on roadside

Table - 1 shows the large variation in the amount of flower wastes generated daily in few famous temples of different parts of India. Trivedi (2006) has reported that temples in Mumbai generated about 15 tons of Nirmalayas waste every day; while the Mahakal temple and others in Ujjain generated 10-15 tons and all those flower wastes were dumped in the nearby Kshipra river or thrown away which usually hurt religious sentiments of the devotees. Over 25,000 marigold garlands are offered daily at the Mahaveer temple near Patna railway station, the demand for marigold garlands goes up to about 52,000 on every Tuesday and Saturday. Over 10 tons of marigold flowers generated by temples in Patna are thrown into the river Ganga (NNE, 2013). The Shree Jagannath temple at Puri also generates 1.23 tons of waste each day. During festive occasions, this quantity shoots up to 3.5 tons/day (Mohapatra, 2010). The famous Hindu temple at Sabarimala, situated on the dense forests of Western Ghats at about 914 metres above the sea level and four km uphill from the Pamba in Pathanamthitta district, was visited by around 15 million pilgrims in November 2011 to January 2012. In a massive clean-up operation initiated by the Mata Amritanandamayi Mutt, volunteers collected about 35,000 big bags of garbage, including clothes and plastic left behind by the pilgrims from the shrine complex from the foothills of the temple and the river.

**Table – 1: Generation of flower waste in different temples of India**

| Temples | Place | Quantity |
|---|---|---|
| Siddhivinayak Temple | Mumbai | 1.2-2.0 q/day |
| Lord Jagannath Temple | Puri | 1.2 – 3.5 t/day |
| Kashi Viswanath Temple | Varanasi | 2.0 – 2.5 t/day |
| Khajrana Ganesh Mandir | Indore | 4.05 t/day |
| Ranjit Hanuman Mandir | | 2.10 t/day |
| Lord Kamtanathji Temple | Ramghat | 5.475 t/year |

Photo - 5: Flower waste thrown in Ganga River

According to Shinde (2007) roughly 4000–6000 devotees visited daily at Tirupati Balaji temples (the number increased to more than 10,000 on peak occasions) and generated about 35 tons wastes per day in Tirumala. Everyday more than 20,000 coconuts were offered to God for puja. In Varanasi, the city of temples, the generation of waste material each day weighing 3.5 to 4 tons in the form of floral remains, incense residues and miscellaneous ritualistic offerings to the deities is a cause of great concern to the municipality since it has no policy for the disposal of tons of temple wastes (Mishra, 2013b). At Kashi Viswanath temple itself over 20 quintal flowers and Belpatras (leaves of Bel) are offered daily, the quantity of flowers increases by four to five times during Monday of Shrawan month and other festival times (TOI, 2009). These flower wastes are disposed of into the river Ganga by temple staff. It has also been estimated that in Delhi alone a minimum of 20,000 kg flowers are thrown in the holy river Yamuna from temples and other places of worship, the quantity multiplies manifolds in the festive season making its disposal difficult (Sat Pal, 2013). In Indore, Khajrana Ganesh temple produces around 4050 kg flower waste daily whereas Ranjit Hanuman Mandir generates around 2100 kg everyday (Choukse, 2011). Singh and Gupta (2011) have recorded that the generation of temple wastes was 7.563 tons / year (floral wastes about 5.475 tons / year and prasadam about 2.088 tons / year) from Lord Kamtanathji temple at Ramghat of Madhya Pradesh; while 2.39 tons / year from Hanuman Dhara, Gupta Godawari and Sati Anusuya, etc. Immersion of idols of God and Goddess also causes dumping of considerable quantities of floral wastes in water bodies. During Ganesh festival in 2011 Samarth Bharat Vyas Peeth, a NGO in Thane had collected 10 tons of floral waste from different immersion sites (Karlikar, 2011). There are about 150 major temples and 700 smaller temples around Lucknow city which altogether generate 500 kg of temple waste everyday (Singh, 2014).

## Harmful Impacts on Environment due to Faulty Disposal of Flower Waste

Dumping of flower waste in open places not only hurt the aesthetic sentiments of devotees but also creates filthy smell and unhygienic environment during its slow decomposition. The rotten flowers release bad breathe to make walking, talking and sitting around impossible. Lots of harmful bacteria, fungus and worms grow which pollutes soil, water and air and spread disease epidemic among the peoples of surrounding area. Dumping of flower wastes in the water of temple ponds or holy rivers pollutes the water where devotees take bath before puja or use the river water for drinking purposes causing spreading of both skin diseases and water born diseases like cholera, typhoid, dysentery and other diarrheal diseases. Nair (2012) reported about highly polluted Pamba or Pampa River, which flows near the foothills of the shrine, and in which devotees of Lord Aayappan take a customary bath before climbing uphill to the Sabarimala temple. Coli form bacteria level in Pampa river water shot up to 29,60,000/100 ml on January 13, 2012 from 950 recorded on November 11, 2011 and the level of oxygen content also dropped sharply due to pollution created by pilgrims. High pollution of both Holy rivers, Ganga and Yamuna is well known to everyone. Several reports are also available on polluted rivers such as Gomti river in Lucknow (Singh, 2014), Kshipra river in Ujjain (Trivedi, 2006), Mandakini river in Chitrakoot (Singh and Gupta, 2011), Tapti or Tapi river in Surat (Ujjania and Multani, 2011), etc. due to regular dumping of temple wastes in rivers. About 35 tons wastes generated per day in Tirumala are dumped openly at a nearby land fill site which poses a high risk of contamination of the surface and underground water in this fragile ecosystem of the hills (Shinde, 2007). Immersion of idols of God in nearby aquatic resources like ponds, lakes and rivers is the usual practice throughout India. Singh (2014) has rightly pointed out that people don't just throw idols and ritual materials like flowers, incense sticks, food, sweets, clothes, etc into the river, they pack everything in polythene bags and flow them into the river, thus causing more problem to manage. After immersion of idols, their colours, chemicals and other components that are used for idol preparation get dissolved and lead to significant changes in the water quality (Dhote *et al.,* 2001). Ujjania and Multani (2011) analysed water samples collected from Tapi river at Aswanikumar idol immerses point, Surat during pre-immerse, immerse and post-immerse periods of Ganesh idol immersion and found that except pH and dissolved oxygen all other parameters like temperature, Biological oxygen demand (BOD), Chemical oxygen demand (COD), total hardness, total alkalinity, total calcium, oil and grease, etc increased in river water during immersion, causing deterioration in water quality (Table - 2). Deterioration of water quality due to idol immersion was also reported earlier by Reddy and Kumar (2001) and Malik *et al.,* (2010). The floating materials released through idol in the river, lake etc. after decomposition result in eutrophication of water.

**Photo-6:** Burning of temple waste

The used flowers are part of the organic waste. Similar to other organic wastes, burning of leaves and flowers of the temples also result in harmful emissions in the form of smoke, suspended particles, carbon monoxide and benzo(a)pyrene, which in turn leads to air pollution (Photo- 6). Besides dry throats and burning eyes those suspended particles cause coughing, wheezing, chest pain, shortness of breath and sometimes long-term respiratory problems.

**Table- 2: Water Quality Parameters at Aswinikumar Ovara**

| Parameters | Pre-immersion | Immersion | Post-immersion |
|---|---|---|---|
| Temperature (°C) | 28.0-31.0 (29.1 ± 0.7)* | 29.0-31.0 (29.8 ± 0.6) | 29.8-30.5 (30.1 ± 0.2) |
| pH | 7.2-7.5 (7.3 ± 0.1) | 6.9-7.6 (7.4 ± 0.2) | 7.0-7.4 (7.3 ± 0.1) |
| Free $CO_2$ (mg/L) | 14.08-19.36 (17.2 ± 1.3) | 15.8-21.1 (18.8 ± 1.6) | 17.6-21.1 (17.6 ± 2.0) |
| Dissolved Oxygen (mg/L) | 6.4-6.8 (6.6 ± 0.1) | 6.0-7.6 (6.7 ± 0.5) | 5.6-6.8 (6.4 ± 0.4) |
| BOD (mg/L) | 3.2-4.8 (4.2 ± 0.3) | 3.6-5.6 (4.8 ± 0.6) | 2.4-2.8 (3.2 ± 0.8) |
| COD (mg/L) | 19.4-41.7 (28.57 ± 4.8) | 40.6-44.0 (42,7 ± 1.1) | 36.6-38.3 (37.3 ± 0.5) |
| Total Hardness (mg/L) | 110.0 -168.0 (131.0 ± 13.4) | 144.0-174.0 (160.0 ± 8.7) | 124.0-170.0 (142.0 ± 14.2) |
| Total Alkalinity (mg/L) | 220-290 (252.5 ± 18.9) | 280.0-350.0 (310.0 ± 20.8) | 260.0-290.0 (267.0 ± 8.8) |
| Total Calcium (mg/L) | 60.0-82.0 (73.5 ± 4.8) | 82.0-96.0 (88.7 ± 4.1) | 76.0-86.0 (80.7 ± 2.9) |
| Oil & Grease (mg/L) | 0.43-0.69 (0.59 ± 0.1) | 0.9-1.3 (1.1 ± 0.1) | 0.7-0.8 (0.8 ± 0.04) |

**The sequence of data shown in table is minimum – maximum (mean ± standard error)*

## Beneficial Utilisation of Flower Wastes

Considering daily visit by pilgrims to 1.08 lakh temples in India and use of a good amount of flowers in different social functions, ceremonies and meetings, a huge quantity of flower wastes is generated daily in India, but unfortunately a lion share of this organic resource materials are either burned or dumped in landfills or aquatic bodies without their beneficial utilisation. Different researchers are trying to utilise this rich source of organic matter for the benefit of mankind and also for saving the environment from above mentioned pollutions.

### i) Agarbatti (Incense Stick) production

Scientists of the Central Institute of Medicinal and Aromatic Plants (CIMAP), Lucknow have developed techniques under a project named 'Mission Sakshama' to use flower wastes of the temples for making herbal incense sticks (Agarbatti) from Marigold flowers and rose water by distilling roses. The female workers are mainly trained to earn extra money by converting flowers into agarbatti. The marigold flowers are dried under the sun and then crushed to turn them into a paste. A little quantity of wood powder is mixed with the paste. Coal powder is used as a coating on the sticks. The herbal scents are used for fragrance. These incense sticks are pure and free from any chemicals and thus not harmful to human beings. On an average 1,500 agarbatti can be made from one kg of raw material (Flower powder and binding materials). The womenfolk of villages near Lucknow are manufacturing agarbatti for last four years by collecting flower wastes from Chandrika Devi temple, Devan Sharif and Mankameshwar temple and supporting their families by this part time job (Pandey, 2014).

### ii) Gulaal (colours for Holi festival) Production

Mishra (2013a) has reported that flowers offered to deities in temples and later dumped into rivers are finding their way into colour sold across 13 cities to celebrate an eco-friendly Holi. Delhi based 'Society for Child Development' has started a project to convert flower wastes of temples into pink and red colour. Flowers after collecting from temples are cut and dried throughout the year. The processing of converting flowers into colours is carried out from November to mid-January and then packaging and delivery are done up to February every year for its use during Holi festival in India. The colours are lab-tested and found safe for use even by children. The Society for Child development has sold about 9000 kg of colour after its production from flowers in 18 centres across Delhi with the help of trained 'Mentally – challenged youth' and 'Destitute women'. The project provides a scope to earn income to women self help groups and mentally – challenged young adults.

### iii) Extraction of Natural Dye

Since some of the synthetic dyes are suspected to release harmful chemicals that are allergic, carcinogenic and detrimental to health, the usage of natural dyes prepared from plants, vegetables, flowers, insects, etc is gaining more and more importance for dyeing and printing of textiles. Natural dyes produce very uncommon, soothing and soft shades in textiles as compared to that of synthetic dyes. Teli *et al.*, (2013) prepared dye from Marigold (*Tagetes erecta L.*) and Hibiscus (*Hibiscus rosa-sinensis*) flowers collected from temples as flower wastes and evaluated their potentials by dyeing 100% cotton fabrics under normal dyeing conditions and on cotton silk blend fabric. They prepared 10% stock solution by boiling 10 g each of dry marigold flower petals and dry pulverized hibiscus flower powder separately in 100 ml of water for 60 minutes. The extracts were filtered and volume was made up to 100 ml and used for dyeing. After mordanting of cotton and cotton/ silk fabrics in Rota dyer machine with Myrobolan (Harada) solution for 1 hr, the fabrics were squeezed and dyed separately with marigold and hibiscus dyes in dye bath for 1 hr and then squeezed and washed with cold water. The colours of dyed samples were evaluated by reflectance method, colour fastness by ISO II method and light fastness by ISO 105/B02 method. The study has shown that hibiscus and marigold flowers which were obtained as temple waste have good colouration properties. The fastness properties were found to be well within the accepted limits. Compared to yellow marigold flowers, orange marigold flowers gave higher depth of shade. Hence, waste flowers of both these types can be used for colouration of cotton and cotton/silk fabrics.

The violet coloured petal part of Saffron flower is thrown as a waste after harvesting the stigma from the flower. Raja *et al.,* (2012) prepared water soluble natural dye by soaking 100 g dried saffron flower powder in 2 litres of water for overnight and then by boiling the extract for 1 hr. The filtered extract was then applied on Pashmina wool at two different pH namely pH 4 - 5 and pH 7 - 8 with and without the use of mordant. They found that saffron flower waste extract was able to dye the Pashmina shawl satisfactorily with very good washing and light fastness. The colours were bright greenish yellow and green at acidic pH while light reddish brown at alkaline condition. Since anthocyanidine constituents present in saffron flower extract has the inherent property of exhibiting bathocromic shift due to the deprotonation of phenolic and carboxylic acid groups during change in pH from acidic to alkaline, they suggested to give a soaping treatment for the saffron dyed pashmina fabrics at acidic pH to maintain the colour during usage. The dyed pashmina fabric at acidic pH without mordant was also showed antimicrobial efficacy against *Staphylococcus aureus*. Hence, dyeing with saffron flower waste can be considered as a commercially viable process of dyeing Pashmina fabrics with antimicrobial efficacy.

### iv) Vermicompost Production

Flower wastes, being organic and biodegradable in nature can be composted by mixing with cow dung or effective microbial culture. There was little doubt about its palatability to epigeic earthworms. For academic interest to know the time needed for vermicomposting along with its quality evaluation and for convincing vermicomposting producers to utilize flower wastes of temples as important resource materials the author conducted following experiment at Defence Research Laboratory, Tezpur in April 2001.

**1. Antirrhinum, 2. Calendula, 3. Candytuft, 4. Dianthus, 5. Marigold, 6. Petunia, 7. Phlox, 8. Poppy, 9. Saliva**

**Photo- 7:** Seasonal flowers used for vermicomposting

### a) Experimental Procedure

During winter months, lot of seasonal flowers were cultivated in office campus for the purpose of beautification. Nine different types of flowers (Photo-7) such as Marigold, Calendula, Dianthus, Petunia, Antirrhinum, Candytuft, Phlox, Poppy and Salvia were collected separately and air dried under a shade. Their scientific names, order and family of each flower plant are presented in Table - 3. Four kg dry flowers of each variety were mixed with 8 kg cow dung and required quantity of water to make the mixture moist. Round heap bed was made on the cemented floor under shade and covered with black polythene sheet for increasing temperature of the mixture and quick decomposition. For each flower three set of beds were prepared and arranged in completely randomised design. Materials of each bed were mixed after 7 days and covered again. After 15 days, material of each bed was remixed, 200 adult earthworms, *Perionyx excavatus* were released on top of each bed and covered with moist hessian cloths. Moisture of beds was maintained at 80 to 85% by sprinkling water as and when required. Earthworms deposited vermicompost on the top as excreta. When the composting material became black coloured, loosely granular structured material up to lower portion of the bed, as observed by physical verification, the material was spread in the shade for bringing down the moisture to about 40% by air drying after hand sorting of earthworms. Vermicompost of each bed was sieved separately through 2 mm sieve and samples were collected for chemical analysis following standard procedures (Bhargava and Raghupathi, 1993).

**Table 3: Scientific Name, Order and Family of Flowers Used in Experiment**

| Sl. No. | Name of flower | Scientific name | Order | Family |
|---|---|---|---|---|
| 1 | Antirrhinum | *Antirrhinum majus* | Lamiales | Veronicaceae |
| 2 | Calendula | *Calendula arvensis* | Asterales | Asteraceae |
| 3 | Candytuft | *Iberis umbellata* | Brassicales | Brassicaceae |
| 4 | Dianthus | *Dianthus barbatus* | Caryophyllales | Caryophyllaceae |
| 5 | Marigold | *Tagetes erecta* | Asterales | Asteraceae |
| 6 | Petunia | *Petunia x hybrid* | Solanales | Solanaceae |
| 7 | Phlox | *Phlox paniculata* | Ericales | Polemoniaceae |
| 8 | Poppy | *Papaver rhoeas* | Ranunculales | Papaveraceae |
| 9 | Salvia | *Salvia splendens* | Lamiales | Lamiaceae |

### b) Quality of Vermicompost

Results in Table 4 showed that about 4 to 5.3 kg vermicompost was produced within one to one and half months from different flowers except marigold and candytuft flowers which took about 2 and 2.5 months time respectively. These vermicompost were neutral to slightly alkaline in nature (pH 6.3 - 8.7), rich in total nitrogen (0.99 -1.88 %), total potassium (1.48 - 2.54 %) and total calcium contents (0.74 - 2.08 %). Total phosphorus varied from 0.51- 0.87 %; while total sulphur content recorded was 0.95 -1.15 %. Total sodium was found 1.55- 2.50 %.

The nutrient status of vermicompost usually depends on the nutrient content of the resource material from which vermicompost has been made. The percentage of total nitrogen, phosphorus and potassium recorded in vermicompost prepared from different flowers were at par with those available in vermicompost prepared from other biodegradable wastes like agricultural wastes, kitchen wastes, terrestrial and aquatic weeds, residential wastes and dry leaves of avenue trees, etc. (Sannigrahi, 2011). The experiment, therefore, proved that flower wastes were not only good resource materials for producing good quality vermicompost but flowers were also palatable food materials to epigeic earthworms.

**Table- 4: Chemical quality of vermicompost prepared from flowers**

| Flowers | Time (mt) | Yield (kg) | pH (1:2) | Total contents (%) | | | | | |
|---|---|---|---|---|---|---|---|---|---|
| | | | | N | P | K | Ca | S | Na |
| Antirrhinum | 1.0 | 5.0 | 8.7 | 1.48 ± 0.11 | 0.70 ± 0.01 | 1.81 ± 0.02 | 1.61 ± 0.04 | 0.96 ± 0.03 | 2.50 ± 0.10 |
| Calendula | 1.5 | 5.3 | 8.3 | 1.24 ± 0.09 | 0.71 ± 0.01 | 1.83 ± 0.02 | 1.47 ± 0.03 | 1.15 ± 0.04 | 2.13 ± 0.07 |
| Candytuft | 2.0 | 5.1 | 6.3 | 1.88 ± 0.07 | 0.64 ± 0.02 | 2.17 ± 0.05 | 1.67 ± 0.05 | 1.09 ± 0.16 | 1.76 ± 0.12 |
| Dianthus | 1.0 | 4.5 | 7.5 | 1.38 ± 0.10 | 0.62 ± 0.01 | 1.48 ± 0.07 | 1.18 ± 0.06 | 1.01 ± 0.13 | 1.55 ± 0.11 |
| Marigold | 2.5 | 5.3 | 8.0 | 1.64 ± 0.11 | 0.87 ± 0.02 | 2.54 ± 0.07 | 2.08 ± 0.05 | 1.09 ± 0.10 | 1.74 ± 0.07 |
| Petunia | 1.0 | 4.0 | 8.5 | 1.38 ± 0.09 | 0.79 ± 0.01 | 1.97 ± 0.04 | 1.57 ± 0.04 | 1.01 ± 0.06 | 1.97 ± 0.08 |
| Phlox | 1.5 | 5.0 | 8.4 | 1.33 ± 0.08 | 0.71 ± 0.07 | 2.39 ± 0.04 | 1.68 ± 0.03 | 1.01 ± 0.19 | 2.32 ± 0.10 |
| Poppy | 1.0 | 4.8 | 7.6 | 1.56 ± 0.08 | 0.64 ± 0.03 | 1.62 ± 0.05 | 1.81 ± 0.03 | 1.03 ± 0.20 | 1.61 ± 0.09 |
| Salvia | 1.5 | 5.0 | 8.1 | 0.99 ± 0.10 | 0.51 ± 0.05 | 1.67 ± 0.07 | 0.74 ± 0.06 | 0.95 ± 0.17 | 1.61 ± 0.08 |
| Range | 1.0 – 2.5 | 4.0 – 5.3 | 6.3 – 8.7 | 0.99 – 1.88 | 0.51 – 0.87 | 1.48 – 2.54 | 0.74 – 2.08 | 0.95 – 1.15 | 1.55 – 2.50 |

Singh *et al.*, (2013a) prepared vermicompost from temple wastes (mainly from *Aegle marmelas* leaves, flowers of *Datura stramonium*, *Tagetes erecta* and *Hibiscus rosa sinensis*), kitchen wastes {mainly fruit and vegetable peels (60-75%), used tea leaves, flour, rice, pasta, bread, noodles, cooked vegetable and potatoes} and yard wastes (mainly dried leaves of *Mangifera indica, Saraka asoka, Syzygium cumini, Tamarindus indica* and various grasses) after mixing with cow dung (waste : cow dung =1:2, w/w) using earthworms, *Eisenia fetida* and observed that the vermicompost of temple waste recorded higher pH, bulk density, particle density, total phosphorus and total potassium as compared to those of vermicompost prepared from kitchen waste and yard waste (Table - 5). The C/N ratio of vermicompost prepared from kitchen waste and temple waste were found less than 20 which indicated faster rate of decomposition and better degree of organic matter stabilisation as compared to other prepared from yard waste. The earthworm biomass in the three vermicompost at 40 days was found in the order of: 4.2 kg in case of Temple waste > 2.3 kg in kitchen waste > 1.9 kg in yard waste. The highest earthworm population in temple waste vermicompost might be attributed to better physical and chemical conditions for growth and multiplications of earthworms.

**Table – 5: Physico-chemical characteristics of three different vermicompost**

| Parameters | Temple waste | Kitchen waste | Yard waste |
|---|---|---|---|
| pH | 7.9 ± 1.3 | 7.7 ± 0.8 | 7.2 ± 0.6 |
| EC ($dsm^{-1}$) | 4.9 ± 0.7 | 3.5 ± 0.7 | 4.2 ± 0.2 |
| Bulk density ($gcc^{-1}$) | 0.7 ± 0.1 | 0.6 ± 0.1 | 0.6 ± 0.1 |
| Particle density ($gcc^{-1}$) | 1.9 ± 0.1 | 1.6 ± 0.4 | 1.4 ± 0.3 |
| Porosity (%) | 70.9 ± 5.2 | 66.6 ± 6.7 | 60.9 ± 4.0 |
| Total N (%) | 0.9 ± 0.1 | 1.2 ± 0.2 | 1.1 ± 0.1 |
| Total P (%) | 1.3 ± 0.3 | 1.2 ± 0.2 | 1.1 ± 0.1 |
| Total K (%) | 0.5 ± 0.1 | 0.3 ± 0.1 | 0.3 ± 0.1 |
| C/N ratio | 19.4 ± 3.6 | 17.6 ± 3.2 | ± 3.8 |

### c) Effect of physical parameters of Flower waste on growth of earthworms

Vermicomposting is a simple biotechnological process of converting biodegradable wastes into organic manure using microorganisms and epigeic earthworms. The process is faster than simple composting since a significant transformation of waste material takes place during passing through the earthworm's gut, whereby the resulting earthworm castings become rich in nutrients, beneficial microbial activity and plant growth regulators. Certain factors like temperature, moisture, pH and particle size of the waste materials play important role on growth of earthworms and rate of vermicomposting. Gurav and Pathade (2011) observed that the maximum biomass of earthworms and cocoon production were recorded at 25$^0$C temperature, 80 % moisture of the bed, 7.0 pH and 1 – 2 mm particle size of material during vermicomposting of flower waste (500 g each) collected from Ganesh temple at Sangli, Maharashtra using *Eudrilus eugeniae* earthworms for up to 5 weeks after mixing with 500 g cattle dung and 1kg biogas digester slurry (Tables - 6 and 7). Beyond 25$^0$C the *Eudrilus eugeniae* earthworms could not survive. They have recorded total nitrogen, phosphorus and potassium content in vermicompost prepared from temple waste (mainly flowers, leaves, fruits, sugar, jiggery, etc) as 1.58%, 0.33% and 0.28% respectively. These findings confirm that the flower waste can be an important resource material for producing good quality organic manure.

Shouche *et al.,* (2011) also prepared vermicompost from fine chopped (1-2.5 cm size) flower waste mixed with cow dung weight wise in different proportions

(100% : 0%, 75% : 25%, 50% : 50%, 25% : 75% and 0% : 100%) using *Eudrilus eugeniae* earthworms and found that moisture fluctuated from 75 – 80% in 100% flower waste; while the average moisture level was reduced to 57 – 68% as the proportion of cattle dung increased in composting mixture. The temperature was generally increased in earlier period of composting i.e. within 2-3 days when it was increased from initially 25⁰C to 45-60⁰C depending upon the composting mixture and then reduced to 25⁰C after a week. The 100% flower waste generally recorded higher temperature which was reduced as the amount of cow dung increased in composting mixture. At the starting time pH was 6.8 in 100% flower waste and 8.01 in 100% cattle dung. The pH values though decreased a little at the initial stage but increased steadily up to 7.5 to 8.5 till the last phase of composting. The 100% flower waste took longer time to decompose probably due to non addition of cow dung favourable for microbial growth and activities while mixture containing 25% flower waste decomposed faster due to faster microbial activity.

**Table-6: Growth of *Eudrilus eugeniae* after 5 weeks at different temperature and moisture level of vermicomposting material (pH 7.0 and particle size 1-2 mm)**

| Sl. No. | Temperature (°C) | Initial average biomass of earthworms (mg) | Biomass (mg) / cocoon number | Moisture level (%) | Initial Average biomass of earthworms (mg) | Biomass (mg) / cocoon number |
|---|---|---|---|---|---|---|
| 1 | 15 | 155 | 208 / 5 | 50 | 194 | 955 / 8 |
| 2 | 20 | 174 | 1395 / 11 | 60 | 187 | 1565 / 9 |
| 3 | 25 | 166 | **1730 / 14** | 70 | 154 | 2495 / 10 |
| 4 | 30 | 158 | -- | 80 | 171 | **3285 / 11** |
| 5 | 35 | 179 | -- | 90 | 179 | 945 / 8 |

**Table-7: Growth of Eudrilus eugeniae after 5 weeks at different pH values and Particle sizes of vermicomposting material (25°C, 80% moisture and particle size 1-2 mm)**

| Sl. No. | pH | Initial average biomass of earthworms (mg) | Biomass (mg) / cocoon number | Particle sizes (mm) | Initial Average biomass of earthworms (mg) | Biomass (mg) / cocoon number |
|---|---|---|---|---|---|---|
| 1 | 5 | 181 | 645 / 9 | 0.5 – 1 | 183 | 1095 / 6 |
| 2 | 6 | 139 | 5265 / 15 | 1 – 2 | 159 | **1835 / 8** |
| 3 | 7 | 159 | **7090 / 18** | 2 – 4 | 176 | 1475 / 9 |
| 4 | 8 | 180 | 6020 / 16 | 5 – 10 | 164 | 715 /12 |
| 5 | 9 | 172 | -- | 10 – 20 | 151 | 525 /10 |

### d) Development of fungal population during composting of flower waste

Different types of microbes like bacteria, fungi and actinomycetes are responsible for decomposition of all organic wastes including flower waste. During pit composting of temple waste collected from Viswanath temple of Banaras Hindu University campus Singh *et al.*, (2013b) isolated 28 temple waste inhabiting fungus by dilution plate technique. According to their occurrence these were divided into three categories (i) Dominant fungi which appeared more than 70% incidence and more in number (ii) Common fungi which appeared less than 70% incidence and (iii) Rare fungi which appeared once or twice during isolation. The species isolated in these three groups are mentioned in Table- 8. Fungal population such as *Alternaria alternata, Fusarium sp.* and *Rhizopus nigricans were* increased at initial stage but at later stages *Aspergillus niger, Aspergillus flavus, Penicillum rubrum* and *Penicillum citrinum* were recorded predominantly at increasing days.

**Table- 8: Fungi isolated from pit composting materials of temple waste**

| Category | Isolated fungal species |
|---|---|
| Group (1) Dominant | *Aspergillus niger, Aspergillus flavus, Mucor racemosus, Rhizopus stolonifer, Rhizopus nigricans* and *Penicillium rubrum* |
| Group (2) Common | *Trichoderma viride, Aspergillus terreus, Cladosporium cladosporioides, Alternaria alternata, Penicillium citrinum* and white sterile mycelium |
| Group (3) Rare | *Aspergillus sulfureus, Aspergillus luchuensis* and pink sterile mycelium |

### e) Impact of biofertilizers on *Eisenia fetida* during flower waste vermicomposting

Earthworms play important role to convert organic wastes into valuable organic manure and prevailing environmental factors such as food, temperature and moisture have a direct impact on their growth and reproduction as well as on the rate of vermicomposting. According to Edwards (1998), the optimum growth of Eisenia fetida in different animal and vegetable waste was at temperatures of 25 - 30°C and at a moisture content range of 75 – 90%. Now-a-days researchers are applying biofertilizers along with organic manure for enriching the manure and increasing the crop production. Kumar et al (2013) reported that mixing of Azospirillum (AP), Phosphobacteria (PB), Blue green algae (BGA) and Rhizobium (RB) separately with flower waste-cow dung mixture (1:1 ratio, w/w) at the rate of 1g per kg of substrate during vermicomposting for 80 days increased the number, body weight and body length of Eisenia fetida (E f) (Table- 9). The earthworm biomass was increased from 7.28 to 16.56 times in the biofertilizers enriched flower waste vermicomposting process of 80 days which might be due to utilization of micro-organisms as food source along with the substrate. The survival and growth of earthworms even in flower waste-cow dung mixture without biofertilizer confirmed that this waste mixture was palatable and good resource material for preparing organic manure by vermicomposting.

**Table- 9: Relative growth rate of *Eisenia fetida* in biofertilizers enriched flower wastes**

| Treatments | Number of *E. fetida* | | Average body weight (g) | | Average Body length (cm) | |
|---|---|---|---|---|---|---|
| | Initial | Final | Initial | Final | Initial | Final |
| Substrate +Ef | 25 | 206.6±1.1 | 8.74±0.05 | 12.48±1.13 | 6.74±1.92 | 6.78±0.08 |
| Subs + Ef + AP | 25 | 336.4±2.1 | 8.55±0.04 | 15.50±0.20 | 5.82±0.15 | 7.96±0.71 |
| Subs + Ef + PB | 25 | 43604±1.7 | 6.69±0.45 | 13.40±2.14 | 6.20±0.19 | 7.24±0.17 |
| Subs + Ef + BGA | 25 | 280.2±0.3 | 6.14±0.32 | 11.70±0.41 | 6.42±1.13 | 7.38±0.40 |
| Subs + Ef + RB | 25 | 431.2±4.9 | 6.28±0.15 | 16.00±1.58 | 6.42±0.51 | 7.38±0.19 |

### f) Impact of flower waste vermicompost on crop growth

The results showed that the quality of vermicompost prepared from flower waste in no way inferior to other vermicompost prepared from different organic wastes. Gurav and Pathade (2011) carried out a pot experiment with flower waste vermicompost to study the growth parameters of rose and *Hibiscus* plants for a period of three months. Addition of 10% vermicompost in soil gave good enhancement of growth in terms of height, flowering time as well as number of flowers as compared to control (no vermicompost) sets. Singh *et al.,* (2013a) recorded higher shoot and root length, number of secondary roots and total biomass of chickpea plant grown for 25 days with addition of 10%, 12.5% and 25% temple waste vermicompost in pot soil than those of kitchen waste vermicompost confirmed that flower waste vermicompost was superior with respect to kitchen waste vermicompost.

### v) Millicompost production

**Photo- 8:** *A. magna* millipede

Similar to epigeic earthworms, millipedes (Photo- 8) also use decaying vegetable matter and mineral soil as food. They are known as macrodetrivores terrestrial arthropods and represented by more than 80,000 species. Each body segment has two pairs of short legs. They neither bite nor pose any danger to human beings. Each earthworm weighs about 1 to 1.5 g eats waste equivalent to its body weight and produces cast about 50% of the waste it consumes in a day, but mature millipedes eat organic waste about five times of their weight. Similar to earthworm, the gut microbes of millipedes favour the degradation of organic matter and recycling of detritus energy too (Alagesan et al., 2003). Ramanathan and Alagesan (2012) prepared both millicompost and vermicompost from flower wastes (discarded garlands made of Jasminum sambac, Calendula officinalis, Gomphrena globosa, Celosia spicata and petals of Nelumbo nucifera) collected from Meenakshi Sundareswarar Temple in Madurai city. The flower waste after drying and shredding was mixed with equal quantity of fresh cow dung in plastic troughs and allowed for pre-digestion by sprinkling of water. After 30 days, the pre-digested substrates (3 kg each) were subjected to millicomposting and vermicomposting by releasing separately 50 numbers each Arthrosphaera magna millipedes collected from the reserve forest of Alagarmalai Hills and 50 numbers each Eisenia fetida respectively. They compared the quality of millicompost with vermicompost after 60 days of bio-composting and observed that the nitrogen, phosphorus, potassium and calcium contents were comparatively higher in millicompost than that of vermicompost (Table -10). In a pot study to find out the effect of compost processed by A. magna and E. fetida on the growth parameters of Capsicum annum they have observed that plant height, number of leaves, area of leaves, fruit number, fruit weight and total chlorophyll content were greater in millicompost than vermicompost. This confirms the rich quality of millicompost is rich in nutrients than vermicompost and opens up a new avenue to utilize flower wastes with higher benefit.

**Table-10: Physico-chemical characteristics of both millicompost and vermicompost**

| Parameters | Waste materials on 1$^{st}$ day | Millicompost on 60$^{th}$ day | Vermicompost on 60$^{th}$ day |
|---|---|---|---|
| pH | 8.2 ± 0.6 | 6.6 ± 0.6 | 6.8 ± 0.5 |
| Total N (%) | 0.81 ± 0.06 | 0.96 ± 0.07 | 0.92 ± 0.07 |
| Total P (%) | 0.52 ± 0.03 | 0.85 ± 0.06 | 0.76 ± 0.06 |
| Available K (%) | 0.54 ± 0.04 | 0.73 ± 0.05 | 0.65 ± 0.05 |
| Available Ca (%) | 1.24 ± 0.09 | 2.78 ± 0.26 | 2.46 ± 0.19 |
| C/N ratio | 36.46 ± 3.2 | 22.35 ± 2.0 | 25.30 ± 2.2 |

## Commercial Composting of Flower Waste

Aerobic composting of temple waste has been started by Mumbai Grahak Panchayat under the leadership of Mrs Prathibha Bawlekar, vice chairperson for the nirmalaya (offerings of marigold, hibiscus, betel nut flower and durva grass) collected from Parleshwar temple since 2002 and Siddhivinayak temple since 2005 (Jamwal, 2005). They mix nirmalya with a culture of bacteria at the rate of 1g/ kg nirmalaya, every other day for about 20 days. It is important to churn the waste regularly, to prevent the formation of leachate and avoid stink. The compost was prepared ready within 35 days. She worked with the Mumbai based Excel Industries Ltd (EIL) to develop a machine that reduces the total volume of waste and reduce the composting time. Now-a-days the floral waste is fed into the machine along with the powder that is used to culture the nirmalaya. By running the machine for only 15 minutes, the waste gets reduced by about 60%. The waste is then heaped and composted within 15 days. Mahakal temple Trust of Ujjain and Kashi Viswanath Temple Trust of Varanasi have approached Excel Industries limited to install the machine for converting temple waste to compost too (Trivedi, 2006).

Indian Grameen Services- another NGO in Indore has started vermicomposting of temple waste collected from Khajrana Ganesh Mandir and Ranjit Hanuman Mandir since September 2011 (Choukse, 2011). The process takes around 80 days for complete conversion of temple waste in to organic manure.

Niklang Mandal- a community based organization (CBO) of Borivali, successfully produces about 4 tons of vermicompost/ month from 30 tons of flowers delivered by Brihanmumbai Municipal Corporation after collecting from nearby temples (Pandit, 2014). The flower waste was left in a pile to decay for over a month with intermittent turning and twisting. They further transferred the material in to vermicomposting pits (sized 4 x 8 x 4 feet), added cow dung and released earthworm. Water was added frequently in the pit to maintain the moisture. Composting was completed within 45 days.

## Conclusion

With an increase in awareness on the benefits of organic farming amongst common masses and the harmful impact of chemical farming with respect to residual toxicity in food materials and environmental pollution, the demand of sustainable organic agriculture is gaining importance. The lack of availability of sufficient quantity of good quality of organic manure is the real constraint to the organic farmer. Time has come when each and every organic waste should be reutilised for making organic manure without wasting these through burning or dumping. Large quantity of flowers is thrown away after offering them to deities or completion of various functions. The article represents detailed information on the flower waste as one of the important and major resource materials which could be utilised in many ways for the benefit of entire human society and sustainable agriculture.

## References

Alagesan, P., Ashokkumar, B., Muthukrishnan, J and Gunasekaran, P. (2003): Isolation and characterization of gut bacteria of the millipede, *Xenobolus carnifex (fabricus)*. Indian J. Microbiol., 43(2): 111-113.

Bhargava, B.S. and Raghupathi, H.B. (1993): Analysis of plant materials for macro and micronutrients. In: *Methods of Analysis of soils, plants, waters and fertilizers* (Ed. H. L.S.Tandon), Fertilizer Development and Consultation Organization, New Delhi. p. 49-82.

Choukse, S. (2011): Converting flower waste into fertilizer. Daily Bhaskar, MP Chhattisgarh, Dec.10th.

Dhote, S., Varghese, B., and Mishra, S.M. (2001): Impact of idol immersion on water quality of twin lakes of Bhopal. Indian J. Environ. Protection,21: 998-1005.

Edwards, C. A. (1998): The use of earthworms in processing organic wastes into plant growth media and animal feed protein. In: *Earthworm Ecology* (Ed. C. A. Edwards), CRC Press, Boca Raton, Florida, p. 327-354.

Gurav, M.V. and Pathade, G. R. (2011): Production of vermicompost from temple waste (Nirmalya): A case study. Universal J. Environ. Res. Tech. 1(2):182-192.

H.T. (2012): Project to turn flower waste into organic fertilisers. Hindustan Times, Kolkata, November 5th.

Jamwal, N. (2005): Holy manure. Down to Earth. November 30th , 2005.

Karlikar, N. (2011): Floral waste put to good use, fertilizer from it distributed among citizens. The Times of India, 12 April, 2011.

Kumar, D. S., Kumar, P. S., Kumar, V. U. and Anbuganapathi, G. (2013): Impact of biofertilizers on growth and reproductive performance of *Eisenia fetida* (Savigny, 1926) during flower waste vermicomposting process. Annual Review & Res. Biol., 3(4): 574-583.

Malik, G. M., Raval, V. H., Zadafiya, S. K. and Patel, A. V. (2010): Idol immersion and physico-chemical properties of South Gujarat Rivers. Current World Environ. 5(1): 173-176.

Mishra, G. (2013a): Temple flowers blossom in eco-friendly gulaal pouches. The Indian Express. March, 21st.

Mishra, N. (2013b): Unholy mess: Temple waste- a concern. The Times of India, Varanasi, Aug.17th..

Mohapatra, P. K. (2010): Environmental Impact of Water and Sanitation Issues.

Nair, G. K. (2012): And quiet flows the Pampa - with filth and coli form.

N. N. E. (2013): Temple to compost, not dump flowers in river. News NET East, Patna, May13th.

Pandey, S. (2014): Women turn waste into wealth. Deccan Herald, 11 May. Pandit, S. (2014): Divine decompose. Dainik Jagaran City plus, February 9th.

Raja, A. S. M., Pareek, P. K., Shakyawar, D. B., Wani, S. A., Nehvi, F. A. and Sofi, A. H. (2012): Extraction of natural dye from saffron flower waste and its application on pashmina fabric. *Adv. Appl. Sci. Res.*, 3(1): 156-161.

Ramanathan, B. and Alagesan, P. (2012): Evaluation of millicompost versus vermicompost. *Current Science*, 103(2): 140-143.

Reddy, V. M. and Kumar, V. A. (2001): Effects of Ganesh Idol Immersion on some water quality parameter of Hussainsagar. *Current Science*, 81: 1412.

Sannigrahi, A. K. (2011): Agriculture and waste management for sustainable future. New Delhi Publishing Agency, Pitampura, New Delhi. p.168.

Sat Pal (2013): Let a thousand flowers bloom. Millennium post. Delhi, May 14th.

Shouche, S., Pandey, A. and Bhati, P. (2011): Astes. *J.Environ. Res. Develop.*, 6(1): 63-68.

Shinde, K. (2007): Pilgrimage and the Environment: Challenges in a Pilgrimage Centre. Current Issues in Tourism, 10(4):343-365.

Singh, A. and Gupta, G. (2011): Generated household and temple waste in Chitrakoot- a pilgrimage point in India: Their management and impact on river Mandakini. *Indian J. Sci. Tech.*, 4(7):750-758.

Singh, A., Jain, A., Sarma, B. K., Abhilash, P. C. and Singh, H. B. (2013a): Solid waste management of Temple floral offerings by vermicomposting using *Eisenia fetida*. Waste Management, 33: 1113-1118.

Singh, P. (2014): Polluting Gomti regularly religiously. The Times of India, Feb.3rd.

S., Bikram, B., Mishra, J., Trivedi, P., Kumar, R. A., Yadav, S. M. and Sinha, A. (2013b): Studies on quantitative and qualitative analysis of fungal population of decomposing temple wastes in Varanasi region. *Plant Pathology J.*, 12(2): 110-114.

Teli, M. D., Valia, S. P. and Kolambkar, D. (2013): Flower waste from temple for dyeing of cotton and cotton/silk. Journal Textile Association. 74(4): 210-214.

TOI (2009): Floral wastes continue to pollute Ganga. The Times of India, Varanasi, July 8th Trivedi, S. (2006): Manure from temple flowers, Excel Industries to invest Rs 5-10 lakh in the venture.

Ujjania, N. C. and Multani, A. A. (2011): Impact of Ganesh Idol Immersion Activities on the Water Quality of Tapi River, Surat (Gujarat). *Indian Res. J. Biol.*, 01(01): 11-15.

# 2

# Fruit and Vegetable Waste: An Easily Available Source for Organic Manure

**S.M. Singh**

*Department of Animal Science, Faculty of Applied Sciences, Mahatma Jyotiba Phule Rohilkhand University, Bareilly (India)*

## Fruit and Vegetable Waste: the Present Scenario

India - the world's second largest producer of fruits and vegetables is throwing away fresh produce worth INR 133 billion every year because of country's lack of adequate cold storage facilities and refrigerated transport. The production of fruits, vegetables and spices was nearly 287 million tons in 2016-17 in India as per the estimates by the Department of Agriculture and Farmers Welfare; the fruit and vegetable sector generates large amounts of waste. In industrialized countries, **fruit and vegetable waste** (FVW or Mandi waste) is mainly generated before reaching consumers, due to programmed overproduction and unfulfillment of retailer quality standards. FVW poses environmental problems due to its high biodegradability, represents a loss of valuable biomass and an economic cost for companies. Although different strategies have been proposed to tackle FVW like reduction, reuse and recycling but successful results have not been achieved even today. Misi and Forster, (2001) have reported that fruit and vegetable wastes (FVW) are produced in large quantities in markets and constitute a source of nuisance in municipal landfills because of their high biodegradability

In India, there are about 400 districts with about 4000 towns and bazaar places having large vegetable markets. The vegetable waste at all these places is nearly around fifty thousand tons / day. At least 20% vegetables and 15% fruits we bring to our shops go waste at the end of the day. These are mostly perishable items like green leaves and sensitive fruit varieties that cannot survive for more than a couple of days.

## Bareilly and its Whole Sale Fruit and Vegetable Mandi

Bareilly city has a big whole sale market (Mandi) of fruits and vegetables. This market is spreaded in approximately 2 acres of land in its central area. It supplies vegetables and fruits in the city and neighboring towns too. In an estimate, more than 500 tons of waste of fruits, vegetables and packing material is generated everyday in the Mandi. This solid waste is either dumped at different places on the roadsides or low land areas and is considered as one of the major threats to the human beings heath. This waste plays an important role in the deterioration of the local environment. The local environment around the Mandi area is in the form of a hospital zone from its eastern side; airport area from the west; two central research institutes from the south and several residential colonies from its north side (see location map, photo- 1&2) The aim of this chapter is to make an attempt to disclose the most effective and low cost methods for its proper and useful transformation in to organic manure.

**Photo-1:** BareillyMandi.

**Photo-2:** Paddy straw as packing waste in Bareilly Mandi

Bareilly-the capital of Rohilkhand region is situated in between Lucknow and New Delhi. It has total population of approximately **4,465,344,** as per the 2011 census. It has been reported that human population is increasing @27% per decade in Bareilly.

## Adverse Effects of Improper Dumping of Fruit and Vegetables Waste

1. The fine organic matter of the waste on dumped sites gets mixed with percolating rainwater –forming leachate. This leachae has high potency to pollute water and sol as well. It causes leaching of NPK and heavy metals like, Pb, Cd, and Cr etc .to deeper soil layers. This results pollution of ground water. It also increases the level of (Total dissolved Solids) TDS and heavy metals. It also causes ammonia toxicity and enrichment of phosphate of surface water.
2. Dumping of solid waste causes soil pollution too, ammonia toxicity to soil, increase in heavy metals in upper soil layers, anaerobic and anoxic decomposition emitting bad odor and spreading of bacterial, viral and parasitic diseases.
3. Burning of such waste contaminates air with dioxins and furans (polychlorinated compounds).Emissions of these gases cause eye burning and irritation, skin diseases, cancer, respiratory problems and many retard tissue growth in brain.

Due to such known adverse effects of solid waste on the human health and the environment, the educated people are now taking deep interest to find out a system of its disposal that could be safe, free from hazards, cost effective and environment friendly. The present topic on Mandi waste (waste of fruits and vegetables) of Bareilly has been chosen for its transformation applying epigeic earthworm, *Eisenia fetida*. The work was conducted at centre of vermiculture and vermicomposting of the university. This technique will reduce the environmental degradation and helpful in protecting the local environment.

## Survey of Mandi : Collection and Segregation of Fruit and Vegetable Wastes

Survey of Mandi was carried out for two days and we discussed about the rate of generation of fruits, vegetables and packing materials per day from the Mandi market with different shopkeepers and selected the hot-spot points for the collection of waste (see photo-3&4).

**Photo-3:** Showing discussion with a fruit merchant

**Photo-4:** Discussing with Mandi Incharge Mr. K. D. Sharma & Mr. Abdul Hussian

Twenty five quintal fruit and vegetable waste (mainly waste of banana packing, orange, tomato, potato, paddy of the straw and mixed waste) was collected from the identified areas of the whole sale Mandi , (see photo-5,6 &7).

**Photo-5**: Waste on a mandi street

**Photo-6:** Oranges as fruit waste in the Mandi

**Photo-7:** Vegetable waste of Mandi along with packing waste

Filled it in a tractor-operated trolley and finally transported to vermiculture and vermicomposting centre of the university (see photo-8). Collected waste was segregated from its non-biodegradable parts like plastic, bottles, stone pieces, rubber parts etc. if these are there.

**Photo-8:** Vermiculture and Vermicomposting Centre of the University

## Bed Formation and Processing of Waste

The segregated fruit, vegetable and packing waste (banana leaves)were cut in to small pieces of approximately 1" of size an all the quintal of mixture of waste spreaded in the form of a layer in 5 Vermicomposting beds, each bed containing 5 quintal of waste. A 6" thick layer of pre-decomposed cattle dung was also made just above the layer of Mandi waste in each bed. Mature earthworms of the genus, Eisenia fetida (see photo-9) were introduced@ 200 worms/sq.m area of the bed. Moisture level of the beds was regularly maintained by sprinkling the tap water if required (see photo-10&11)

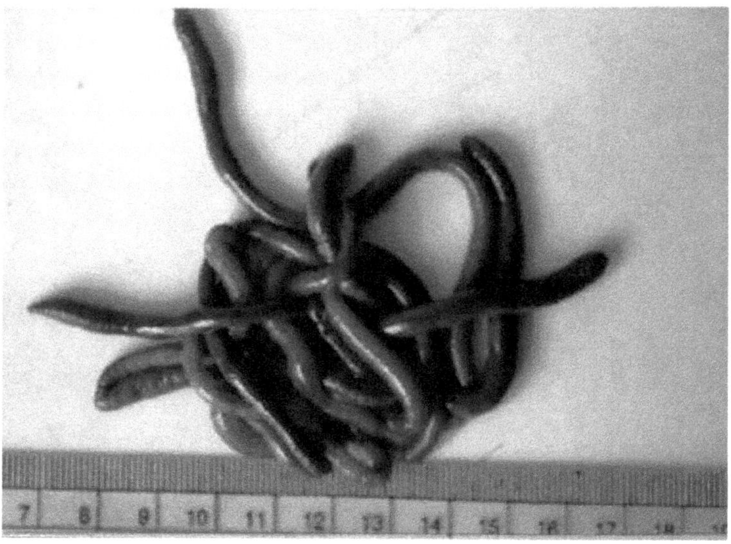

**Photo-9:** Earthworms applied for vermicomposting

**Photo-10:** A bed of Vermiculture & Vermicomposting Centre showing Vermicomposting process

**Photo-11:** Vermicomposting bed showing proper level of moisture content

The bed material was turned up and down every 15 days of time for two times. It may be seen that the color of the waste gets darkened slowly and slowly and the number of the earthworms gets multiplied during the process of vermicomposting. After 45 days of time the waste material along with the cattle dung gets transformed in to a black tea color tea like granulated powdery structure. At this stage, three cone like structures (one in the centre and two on either sides of the bed) were formed in each bed. These cones were left for 24 hrs. During this time, worms moved to the bottom of the cones. On the next day, vermicompost was taken up from the top of the cones up to 3" from the bottom surface. Collected vermicompost was filtered by using a sieve of 3mm size. The physico chemical parameters of mixed waste and Vermicompost were measured before packing.

**Analysis of waste and organic manure**

| Initial parameters of waste (1:1) | Parameters measured | Final Parameters |
|---|---|---|
| 67.3 | % MC | 53.3 |
| 08.4 | $P^H$ | 7.2 |
| 40.3 | %OM | 20.2 |
| 1.3 | %N | 2.3 |
| 23.3 | %C | 11.7 |
| 17.9 | C:N ratio | 4.3 |
| 0.20 | % P | 0.35 |
| 1.72 | %K | 2.46 |
| 1.02 | Zn (ppm) | 1.89 |
| 24.9 | Fe (ppm) | 28.06 |
| 0.24 | Cu (ppm) | 0.30 |
| 3.62 | Mn (ppm) | 4.19 |

## Organic Manure (Vermicompost), its Quality and Utility

Vermicompost- a transformed waste after the egestion process of the worms is highly beneficial to the soil (see photo-12). it can be used in soil as it improves structural stability of the soil stimulating the formation of soil aggregates through bacterial polysaccharides by enhancing the rate of organic matter decomposition(Reddy & pattnaik,2009).It also reduces the C/N ratio in soil-an important factor for better growth and development of the plants.

It can be used as Fasal Sudha (organic fertilizer) as it contains all the micronutrients like, Zn, Cu, Mg, Mn, S along with all the major macronutrients (NPK) in appropriate amount and therefore this vermicompost supplies all the essential dietary substances to the soil system and finally to the plants.

Edwards (1998) has mentioned that the vermicompost is an ideal organic manure for the boosting the plant growth and enhancing the biomass production of wide variety of crops.

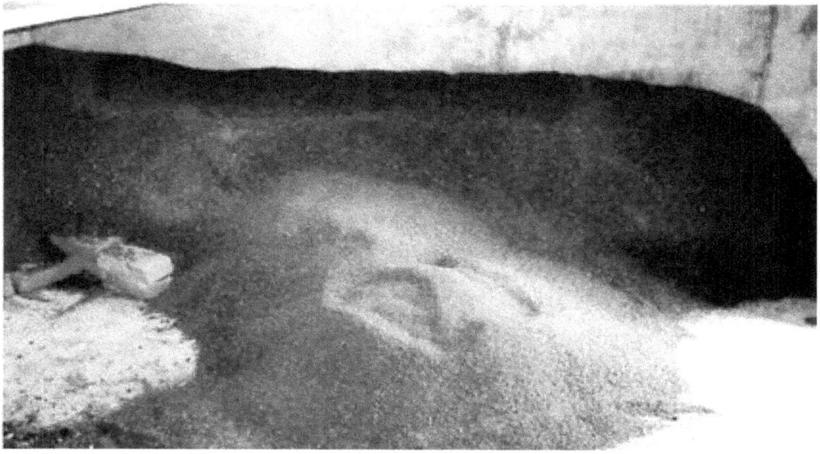

**Photo-12:** Ready Vermicompost from the vegetable and fruit waste of the Mandi

## Findings and Conclusions

It may be noticed that out of 25 Q mandi waste and 25 Q cattle dung (food base of earthworms) 45 Q vermicompost was prepared only in the 45 days of time. In other words, 90% biomanure was formed using vermicomposting technique. pH of the vermicompost was recorded 7.2, nearly natural, which is beneficial for the yield of the crops.

It may be concluded that the fruit, vegetable and its packing material may easily be transformed in to vermicompost, which is highly beneficial for better and sustainable growth of plants and crops. It has been reported that vermicompost promotes growth of a wide variety of cereals, vegetables, fruit plants ornamental plants (KALE, 2011), while working on the kitchen waste also mentioned that the vermicompost is very useful for the plant growth. Umamaheswari and Vijaylakshmi,(2003) have mentioned that the vermicompost prepared from the composting of paper mill sludge using native earthworm species has a good impact on the growth and yield of black gram. Besides above-mentioned qualities, the other benefits of using this technology are:

- It minimizes the amount of solid waste, so reducing the level of local environmental pollution and its hazards.
- It reduces total environmental pollution that may be of air, water or land.
- It is less labour oriented.
- It is job oriented.
- Expenditure is the minimum and benefits are more.
- Earthworms ones purchased, need not to purchase again as their number always increases during the process of vermicomposting.

- It is a profitable business.
- The excess of earthworms may be sale out and get the money.
- Excess of vermicompost can be given to other farmers @ of Rs.4/kg.
- The vermicompost boost up the fertility status of the farming land which has been lost by using excessive chemical fertilizers and pesticides.
- We can move again towards 2nd green revolution and save the foreign exchange by using this organic manure.

## References

EDWARDS, C.A. (1998): The use of earthworms in the breakdown and management of organic wastes. In: C.A. Edwards (Ed.), *Earthworm Ecology,* CRC Press, Boca Raton, FL, USA, pp.327-354.

KALE, R. D. (2011): Origins and spread of vermicomposting in India: Focus on sutainable agriculture. In: Vermiculture Technology. Earthworms,Organic wastes and Environmental Management, (Eds. C. A. Edwards, N.Q. Arancon and R. Sherman), CRC Press. Florida, USA, pp. 469-480.

PATTNAIK, S., AND REDDY, M.V. (2009) : Bioconversion of municipal (organic) solid waste into nutrient rich vermicompost by earthworms (*Eudrilus eugeniae, Eisenia fetida and Perionyx excavatus*). Dynamic Soil, Dynamic Plant 3 (Special issue 2), 122-128.

MISI, S. N., and C. F. FOSTER. (2001): Batch co-digestion of multicomponents agro-wastes. *Bioresource Technology*. 80:19–28. 10.1016/S0960-8524(01)00078-5.

REDDY, M.V., AND PATTNAIK, S. (2009): Vermicomposting of Municipal (Organic) Solid Waste and its implications (Chapter 10). In: Singh SM (Ed.), 'Earthworm Ecology and Environment', International Book Distributing Co., Lucknow India, p. 119-133.

UMAMAHESWARI, S., and G.S.VIJAYALAKSHMI. (2003): Vermicomposting of paper mill sludge using an African earthworm species, *Eudrilus eugeniae* Kinberg with a note on its physicochemical features. *Poll. Res.* 22: 339-341.

# 3

# Paper Mill Sludge and Waste Paper-Promising Sources of High Quality Vermicompost as Organic Manure

*O.P. Agrawal, Puneeta Dandotiya and Muddasir Basheer*

*Vermibiotechnology Centre, School of Studies in Zoology, Jiwaji University, Gwalior - 474011*

## Importance of Paper in the Human Society

Discovery of paper is a significant event in the human civilization. Paper is widely used in day-to-day human life, in academic institutions, offices, business concerns, packaging industries and at homes. Paper sheets are used for writing, typing and printing. Different types, colors and grades of paper are made for different purposes. Paper cardboards are extensively used as packing material (cartons). Paper shredding is used for packaging and transport of delicate, expensive and breakable items, fruits, vegetables, foods etc. It is also used to prepare disposable items like plates, glasses, bowls etc. for catering food and drinks in parties and elsewhere. Paper has played and still playing a leading role in education, creation of knowledge and mass communication. If there was no paper, perhaps the knowledge could not have been preserved and presented in the form of books and other writing and printing formats. Can anybody think about newspapers without paper? Even today, when concept of *e-newspaper,* is available, print version of newspaper is an essential and integral part of life. Many people do not enjoy their morning tea or coffee without newspaper.

## Disposal and Management of Paper Waste: Environmental Concerns

The pulp and paper industry is one of biggest industrial sectors with regards to its significance to the human society as well as to environmental pollution through waste generation. A huge amount of water is consumed in the production of paper. Approximately 250-350 $m^3$ of water is consumed per ton of paper and about 90% of the consumed water (225-275 $m^3$) becomes polluted and turned into waste water.

The treatment of waste water from paper industry, involving physical and chemical methods and is very expensive (Gohil, 1995). In addition to it, a large volume of sludge (press mud) is also produced out (Raghuveer and Sastry, 1995). It is usually flushed off in the water bodies resulting in water pollution and increasing BOD. According to an estimate, the global paper industry is growing at the rate of 2.8% per annum and generation of paper waste is rising too. Thus more efforts will be needed for disposal and management of industrial and used paper wastes and more space will be required for dumping and land-filling of such wastes.

In public and private sectors as well as in educational institutions, the confidential documents like paper sheets, registers and notebooks are not sold in the market due to risk of leakage of secrecy. These are torn into pieces or burnt. The paper and card board wastes from homes and market are also not sorted out for recycling. A huge amount of such paper waste, along with other waste is disposed of by dumping, land-filling and burning. These conventional methods of disposal and management of waste are not satisfactory and are major causes of environmental pollution and hazards to human health.

## Necessity of Recycling of Paper Waste

Proper disposal and management of paper factory sludge and paper waste are obligatory issues to minimize the pollution caused by them. Although the paper sludge appears to be a good source of organic carbon, but it is deficient in other nutrients, hence it is not suitable to be used as source nutrients in the soil. Appropriate recycling of these wastes can generate some useful end products. The best solution of waste disposal is to opt 'reduce-recycle-reuse' technology for paper waste.

## Ways and Means of Recycling of Paper Waste

The waste stuffs from paper factory is managed like other waste stuffs including dumping, land-filling, burning and flushing in the water bodies. These methods are not environmentally safe and are the core causes of soil, water and air pollution and hazardous to environmental and human health. Paper is a natural biological substance made from wood pulp. Its major constituents are cellulose and lignin which are degraded very slowly by the micro-organisms. The structural polysaccharides of paper sludge and paper are extremely stable due to slow biodegradability (Adney *et al.,* 1991). Earthworms are not able to digest paper and other cellulose and lignin containing waste materials by their own digestive enzymes. It is rather not easy to use it for composting and vermicomposting. High content of toxic residues and heavy metals in paper mill waste water and sludge also aggravate the situation. It is because of this reason the paper sludge is sometimes dehydrated to convert it into cake for disposal in land-filling sites and incineration. These methods are not environment friendly and involved a great loss of valuable cellulosic resources.

Suriyanarayanan *et al.* (2010) have carried out a study on the characterization and possible re-utilization of solid wastes from paper industry. They have suggested several usage of dried sludge from a recycled paper mill: as fuel or energy source in bark boilers, as bulking agent or carbon source for composting, as animal bedding, as raw material in the manufacturing of absorbents, as alternative cover at a sanitary landfill along with soil in 50:50 ratio and as raw material in the manufacturing of particle board.

The situation of disposal and management of paper waste produced by households and offices is not better in India and elsewhere. The paper waste is disposed in trash bins along with other types of solid and semi-solid wastes. A major part of municipal solid waste (MSW) is due to paper waste that is typically piled up by the side of households, agricultural fields and highways posing serious problems of environmental pollution. In academic organizations like universities and research institutes, paper waste is periodically piled up and set on fire, causing air pollution and wastage of otherwise utilizable cellulose contents. However, some bigger institutions which have huge output of waste paper (few tons/ week) send the waste for reuse in kraft paper mills.

## Management of Paper Waste through Vermicomposting

Vermicomposting is emerging out as a promising tool for effective and eco-friendly management of different types of organic wastes providing a useful end product- the vermicompost which can help in organic farming and sustainable development (Edwards, 1988; Agrawal, 2005, 2009). The vermicomposting involves combined action of gut enzymes of earthworms and micro-flora present in the waste and intestine of the worms (Senapati, 1992; Bhawalkar, 1993; Ghatnekar *et al.*, 1998; Talashikar and Powar, 1998). The paper and paper based wastes are rich in cellulose (carbon) and poor in nitrogen content. They have much higher C/N ratio. Shredded paper has high-bulk density permitting enough aeration, but water soaked stuff becomes too compact and stuffy and does not allow free aeration. The aerobic bacteria are unable to utilize complex cellulose of paper waste. Thus, vermicomposting of paper waste is not an easy task. However, a number of studies have been carried out on the bio-conversion of paper mill and paper based waste into vermicompost (Satchell and Martin, 1984; Edwards, 1988: Butt,1993; Butt *et al.*,1995; Elivira *et al.* ,1995, 1996a,b, 1998; Elvira *et al.*, 1996a; Elvira *et al.*,1996b; Elvira *et al.*, 1998; Gratelly *et al.*, 1996; Ceccanti and Masciandaro,1999; Deolalikar *et al.*,2005; Nath and Deb, 2008; Umamaheshwari *et al.*, 2009; Kaur *et al.*, 2010; Gajalakshmi *et al.*, 2001a; Gajalakshmi *et al.*, 2001b; Gajalakshmi *et al.*,2002; Gajalakshmi *et al.*,2003; Gajalakshmi and Abbasi, 2003; Gajalakshmi and Abbasi,2004; Vasanthi *et al.*, 2011; Mupondi *et al.*, 2010,11).

The studies revealed that paper sludge and used paper waste are rich in carbon containing molecules like cellulose, hemi-cellulose and lignin, have high C/N ratio and are deficient in nitrogenous substances and they need to be spiked by proteinaceous waste for successful composting and vermicomposting. These additives are cattle dung (dairy manure), sewage sludge, dairy waste sludge, baker's yeast. A variety of such wastes have been tried by different workers using different species of earthworms and vivid systems (*in situ, ex situ*, container units) for vermicomposting.

**Pre-requisites for paper waste vermicomposting**

Paper waste vermicomposting requires amalgamation of suitable waste material (s) in proper ratio. This blending will provide proper pH, C/N ratio, bulk density for aeration and water retention capacity (absorbability) of the medium. Another vital necessity is the selection of suitable species of epigeic earthworms for successful and fast bio-conversion. We have to select those worm species which have fast feeding rates. Prior to release of earthworms, pre-decomposting of waste mixture is a necessary step during which it is heated up under the influence of thermophilic bacteria and release of toxic and obnoxious substances as gases. This bio process is followed by reduction of temperature and activation of mesophilic microorganisms (bacteria and fungi) which degrade the waste to make it suitable for earthworms' feed. These worms feed upon and transform it as high quality organic manure.

In spite of significant research on paper mill sludge and waste paper composting, an economic and feasible model has yet to be developed. Most of the research work was performed in plastic containers under laboratory conditions. The actual vermicomposting should be done outdoor in the field conditions as vermibeds in pits, tanks or directly on earth surface as windrows. It should be promoted by the establishment of model vermicomposting units and Vermicomposting Centers for live demonstrations as promotional activities to enhance eco-friendly means of solid waste management. In practice, none of the units are using large amounts of paper waste for vermicomposting. Looking in to the need, an easy technology has been developed to generate organic manure on a large scale as vermicompost from the paper waste for agrarians. It will assist in improving the soil sustainability and lowering the environmental pollution.

**Paper waste Vermicomposting**

An easily operative paper waste technique was developed at Vermicomposting Centre of Jiwaji University, Gwalior in 2005. This centre is serving as an ideal model system both for awareness, training, research and promotional activities in the field of vermicomposting. A large amount of waste paper of Jiwaji University and its different Schools of Studies was collected and utilized.

The paper waste was procured and passed through paper shredder (Fig. 1). The earthworms (*Eudrilus eugeniae*) were taken from mother culture tank situated at

Vermicomposting Center of the university (Fig. 2). A series of experiments were conducted in plastic containers (Fig. 3) using different ratios of paper waste and cattle dung. Four holes were punched at bottom of the containers for drainage of excess of water. The waste was pre-decomposed for a period of two weeks to remove ammonia and other toxic effluents. The heat generated during initial period of waste biomass decay is sub-sided. Twenty five adult earthworms, *Eudrilus eugeniae* were released in each of the containers which were covered by garden mesh. These vermireactors were maintained for a period of 65 days with periodical sprinkling of water. It was noticed that paper shredding alone was not suitable for activities and survival of earthworms. This appears to be due to deficiency of nitrogen containing nutrients, high C/N ratio and inability of microorganisms and earthworms to utilize carbon rich cellulosic substrates. Results obtained for number, weight and biomass production were similar to those of other workers (Kale *et al.*, 1986; Nagavallemma *et al.*, 2004) who have reported increased number and weight of earthworms on the basis of quality and quantity of available food.

The findings of the study revealed that equal amounts (1:1) of shredded paper and cattle dung provided best results with regards to growth of earthworms as well as the degree of composting than other combinations containing higher amounts (2:1 and 3:1) of paper waste (Basheer and Agrawal, 2013). The summary of results in 1:1 combination of two waste mixtures is depicted in Table 1.

**Table-1: Physico-chemical and earthworm parameters of paper waste: dung (1:1) mixture during vermicomposting process**

| S.N. | Parameters | Duration of vermicomposting (in days) | | |
|---|---|---|---|---|
| | | 0 | 30 | 65 |
| 1. | pH | 8.2 | 7.6 | 7.2 |
| 2. | Total nitrogen (%) | 0.15 | 0.23 | 0.36 |
| 3. | Available phosphorus (%) | 0.78 | 0.95 | 1.18 |
| 4. | Exchangeable potassium (%) | 0.090 | 0.132 | 0.36 |
| 5. | C : N ratio | 30.9 | 16.46 | 5.49 |
| 6. | Percent increase in bio-number | - | - | 675 |
| 7. | Percent increase in biomass | | - | 107 |
| 8. | Percent yield of vermicompost | - | - | 48 |

**Fig. 1:** Paper sheredder machine and basket of shredded paper

**Fig. 2:** Mother culture vermicompositing tank Bunch of earthworms in inset

**Fig. 3**: A series of experiments were conducted in plastic containers

In these studies, it may be noted that still dung was better medium than 1:1 combination. It appears that the type of microorganisms and their populations present in the waste mixture were not efficient enough to digest cellulosic wastes rapidly. Therefore a detailed study was further extended to recognize the influence of three additives on paper waste vermicomposting which could provide a good source of microorganisms to hasten the vermicomposting process.

A free choice experiment was conducted in ceramic tank (55x40x15 cm) divided into four equal sized chambers with the help of thermacol sheets having few holes so that earthworms could pass through from one chamber to another, according to their preferential habits (Fig. 4). The first chamber (A) was filled with a mixture of dung and shredded paper (1:1), pre-decomposed for 2 weeks by using water only. Dung and shredded paper were filled in chamber B, which was pre-decomposed by sprinkling with a solution containing butter milk and jaggery (450 ml + 250 g in 5 litre of water). In the chamber (C), mixture of dung and shredded paper was filled which was pre-decomposed by adding vermiwash and in the chamber (D), mixture of dung and shredded paper was filled, which was pre-decomposed with a solution containing 5 g of *Trichoderma harzianum* in five litre of water. The quantity of the medium in all the four chambers was kept constant (3 kg.). 100 adult earthworms were released and the whole assembly was covered by garden mesh net. Free choice experiment was repeated three times and the results were recorded after 15 days by counting the number of earthworms.

**Fig. 4:** Free choice experiment

Results of free choice experiment showed that the maximum number (35 %) of earthworms preferred *Trichoderma* treated medium for their settlement followed by vermiwash (28 %), jaggery and buttermilk (21%) treated media. The least preference was shown for control (water treated) (17 %) medium (Fig. 5). These findings indicated that the preference might be related with microorganism based enhancement of decomposition of the substrate. In order to confirm this interpretation, another experiment was conducted in plastic containers.

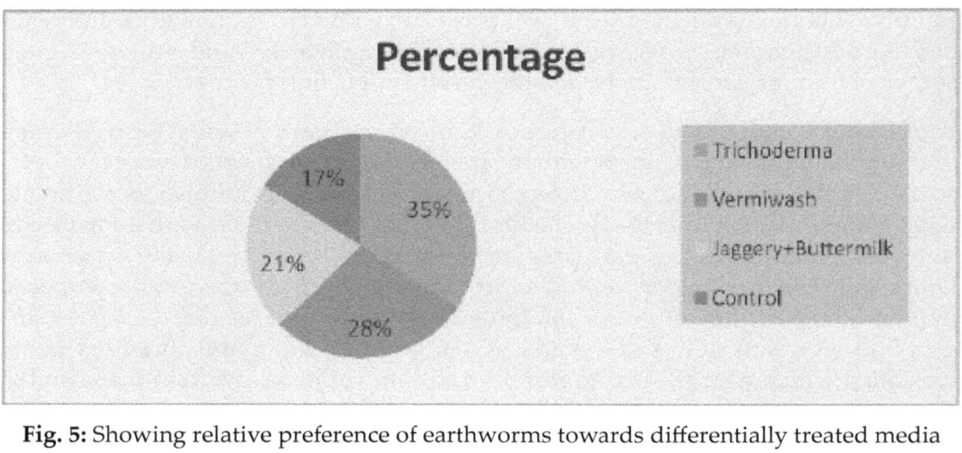


**Fig. 5:** Showing relative preference of earthworms towards differentially treated media

A fixed amount (3 kg) of substrate mixture (shredded paper: cattle dung: 1:1) was filled in plastic containers and the mixtures were separately treated with three selected additives and also with water (control) during pre-decomposting period of 2 weeks as done in "free choice" experiment. Then 25 adult worms were released and observations were recorded after 8 weeks. Each treatment was replicated thrice. The number and weight of adult, baby and juvenile worms and cocoons were recorded and the data were used for calculation of percent increase in bio-number (=population growth) and biomass production. Better results were obtained in substrates treated with all the additives as compared to control (no microbial inoculation). It was noticed that best results were depicted by *Trichoderma* treated substrate followed by vermiwash, jaggery+buttermilk and control. These results are in accordance with free choice experiment. Finally the substrate was air dried and filtered to obtain vermicompost that was weighed and the degree of composting was calculated. Degree of composting was observed the maximum in *Trichoderma* treated medium (59 %), followed by vermiwash (50 %), Jaggery + Buttermilk (45 %) and water control (42 %) treated substrate. In the studies, a system of percentile scoring was used, the final values of percent population growth, earthworm biomass produced and degree of composting were summed up and percentile scores from the maximum (100 %) to the minimum were calculated (Fig. 6).

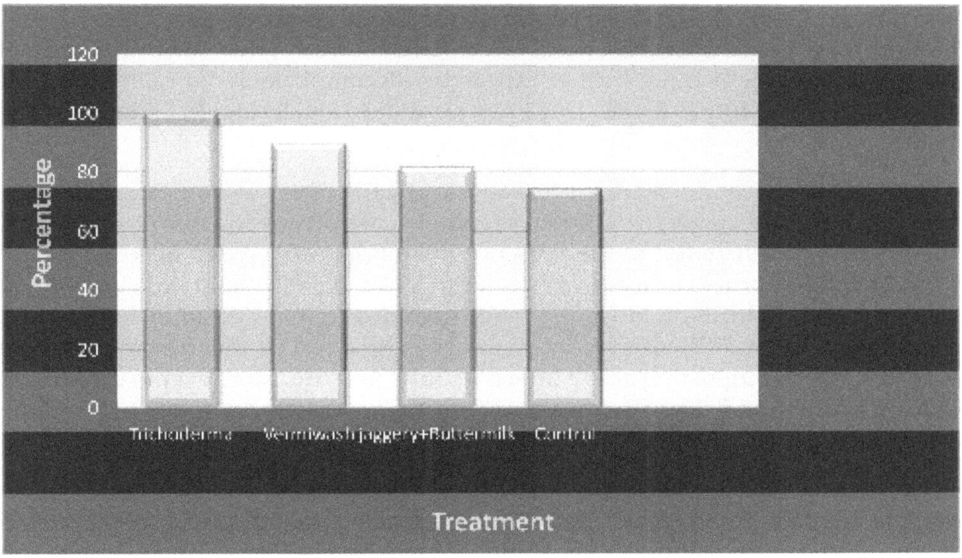


**Fig. 6:** Showing the net percentile score of Vermicomposting of different media treated with different additives

It may be concluded that waste paper generated from anthropogenic sources which is usually burnt or dumped off can be recycled by mixing with cattle dung through vermicomposting and the process can be activated by pre-treatment with *Trichoderma* powder. As a cheaper substitute of such microbial inoculation (*Trichoderma*), vermiwash may be employed that can easily be produced in vermicomposting unit itself and its results were only marginally lower than *Trichoderma*.*

**Trichoderma viridae* and *T. harzianum* are the powders of fungal species of agricultural importance. These are applied in agricultural fields to restrict growth of other harmful fungi in the soil. These have already been used as a favourable bio-inoculants in a number of studies on vermicomposting where they restrict growth of micro-flora and accelerate biomass degradation during pre-decomposition process through production of cellulose, hemicelluloses and lignin degrading enzymes. Rasal *et al.* (1988) reported rapid decomposition of sugarcane trash with a mixture of cellulolytic fungi, *T. viridae, T. spiralis, Paecelomyces fusisporus* and *Aspergillus* sp. along with nitrogen fixing bacteria, *Azotobacter* sp. *Basidomycetes* are the best known lignin degraders and *Pleurotus sajorcaju* has varying enzyme activities and *Trichoderma* and *Aspergillus* can degrade hemicellulose and cellulose respectively (Buswell and Chang, 1994; Milala *et al.*, 2009).

Vermiwash has never been used as stimulant for vermicomposting. It is a rich source of hydrolytic enzymes and is believed to help earthworms in dissolving all kinds of substrates and to protect them from pathogenic microbial infections. It selectively kills harmful microorganisms and favours beneficial ones (Pramanik and Chung, 2011). It has been recommended that vermiwash is a good liquid bio-fertilizer and a microbial pesticide as it includes a good deal of beneficial microorganisms which can further help in bio-decomposition of organic waste in the soil. In these studies it was used as an additive considering that it may help in biodegradation of complex ingredients of paper waste.

There are some recommendations for routine practice of vermicomposting. A spray of buttermilk and jaggery solution can enhance the process of vermicomposting. Its influence is attributed to the presence microorganisms, yeast and lactobacilli in fermentation process. However, Ravikumar *et al.* (2009) could not observe any influence of buttermilk and jaggery treatment during pre-decomposting period. Similarly Parray (2012) also cold not observe any effect of this treatment on his study on vermicomposting of leaf litter and dung mixture. More ever in the present study a positive and slow effect was seen during pre-decomposting of paper-dung waste mixture.

A number of workers have reported that decaying food and vegetable wastes are the sources of pollution, un-hygienic atmosphere and spread of infectious diseases. Due to decaying process, the food waste gets heated up and becomes soft and pulpy with oozing of liquefied lechate. This lechate is the source of pollution, foul smell and it acts as a medium for breeding of anaerobic and pathogenic microorganisms and vectors of diseases. An attempt has also been made by us to develop a simple and safe method for vermicomposting of household (food and kitchen) waste.

It was demonstrated by Sunitha *et al.* (2012) that if kitchen refuses are mixed up with other suitable waste stuffs like cocopith, bagasse or jute waste, problems of bad smells, pollution and un-hygienic atmosphere due to lechate can be resolved. The added stuff acts as lechate absorbing raw material (LARM) and it also serves the purpose of balancing the nutrient content, C/N ratio and bulk density of the waste mixture. The mixture becomes suitable for complete aerobic composting and vermicomposting. A similar approach was followed by Dandotiya and Agrawal, (2013, 2014) on vermicomposting of food waste from market and homes (Fig. 7). In these studies we have included shredded paper sand-soil (1:1) mixture which was supposed to act as lechate absorbing raw material (LARM) from highly nutritious decaying food waste. Experiments were conducted in plastic containers using different combinations of three waste materials, food waste, shredded paper and sand-soil (1:1) mixture. It was observed that the lechate fluids released by food waste are absorbed by other two wastes and heat generation and foul smell was reduced to a great extent. Sand-soil mixture and shredded waste paper also act in balancing the nutrient content, C/N ratio and bulk density of the waste medium. After pre-decomposition period of 20 days, 30 earthworms were released in each container. The cultures were maintained for three months and observations were recorded on number and weight of worms, temperature, pH and NPK value of the medium. Different results were obtained in different combinations of three waste mixtures. Dung slurry (50%) was sprinkled for maintaining proper moisture in the vermicomposting medium. The best results were obtained in 5:1:2 mixture of vegetable waste, paper waste and soil-sand mixture with a maximum increase in worm population (664.43 %) and worm biomass (395.40 %). Analysis of vermicompost revealed maximum nitrogen (1.94%), potassium (1.19%) and phosphorus (1.11%) content in this mixture. It is interesting to note that vermicomposting performance of vegetable waste with paper waste and sand-soil (5:1:2) mixture was at par with dung alone with little use of dung slurry. It is therefore, concluded that vegetable waste can easily be vermicomposted at home level without involvement of complex and cumbersome processing and without polluting the environment.

Fig. 7 — Different types of food and vegetable waste used for vermicomposting

In further studies, it was noticed that amount of sand and soil mixture can be reduced and the same amount of vermicompost can be mixed in food and shredded paper waste. Addition of vermicompost was found to enhance the process. It appears to work like microbial inoculation for faster degradation of cellulose content of the paper waste. It was also noticed that if dung is not available simply water can be used for maintenance of moisture. Plant waste from house gardens and potted plants can also be included in the waste mixture.

## Model of Paper Waste Vermicomposting Unit

Based on the findings of vermicomposting experiments conducted in plastic containers, a model of fully operational vermicomposting tank (8X8X2′) was established. The tank was divided into four equal sized chambers by using bricks (Fig.8). Enough holes are left in the brick partition so that earthworms have a freedom to move from one chamber to other. The first chamber was first filled with a mixture of food waste, shredded paper and vermicompost (5:1:2). The mixture was left for a period of 15 days for pre-decomposting with occasional sprinkling of water or dung slurry as and when required.

**Fig. 8:** Four Chambered tank used for paper waste vermicopositing

It was observed that the temperature of mixture rises up with emission of foul smell. But due to presence of vermicompost and shredded paper, sludge formation by lechate of food waste was prevented, temperature rise was not very high and the foul smell does not last for a longer period. Due to aerobic conditions and presence of enough moisture, the temperature becomes almost normal in about 10 days. After 15 days, some space was made in one of the corners of the chamber and approximately 5 kg of mother culture (4 kg vermicompost containing 1 kg of earthworms) was filled in this space. The earthworms soon get moved in the whole waste stuff. About 20 days after onset of the experiment, the second chamber, in clock-wise direction, was filled with waste mixture like the first one. It was maintained as earlier manner. When the mixture becomes ready for feeding, earthworms migrate into it. Depletion of nutrients and increasing population of worms also facilitate movement of earthworms.

In a similar way third chamber was charged after next 20 days. When the time comes for filling up of the fourth chamber, more or less the whole waste stuff in first chamber was transformed into vermicompost. By this time most of the earthworms also moved to the adjoining chambers. Vermicompost was now ready to be harvested (removed, air dried and sieved). The chamber was now ready to be charged again with the mixture of 3 waste stuffs for next cycle of vermicomposting. It may be required to remove some of the earthworms, when their population becomes too high due to continuous reproduction in suitable atmosphere and management conditions. The productivity of 4-chamber system (vermireactor), in terms of biomass and vermicompost production, was much better than single chamber vermireactor. It works like an automatic system and requires much less manpower in the segregation of earthworms, harvesting and sieving of vermicompost.

This model of food and kitchen waste spiked with paper waste can be adopted as home vermicomposting unit and majority of household wastes (paper, garden, food and kitchen waste) can be processed at source of origin and this will reduce the burden of transport and management of waste of town and city administration (Nagar Palika and Nagar Nigam). Such a system can be employed for vermicomposting of all kinds of waste including leaf litter, cattle dung, flower (temple) wastes etc.

## Prospects of Paper Waste as Source of High Quality Vermicompost for Organic Farming

Paper is an essential commodity to human society. It is used in a variety of ways in academic institutions, offices, business concerns, packaging industries and at homes. A huge amount of paper waste is produced by households, offices and industries. It eventually adds up to the large quantities of municipal solid waste (MSW) that is typically piled up on land by the side of households, agricultural fields, and highways, posing serious problems of environmental pollution. For secrecy reasons, paper waste is not sold in the market and usually torn into pieces and burnt. At many places paper waste is periodically piled up and set on fire, causing air pollution and also wastage of otherwise utilizable cellulose contained in the paper. Only a small part of paper waste is sold to the venders who send it to the paper mills for recycling. The paper and card board waste from market yard is not sorted out for recycling. Due to deficiency of nutrients and very high C/N ratio, paper waste alone is not suitable for composting and vermicomposting. It can be mixed with other wastes to convert it into high quality vermicompost as organic manure.

The present model of paper waste vermicomposting provides an easy, eco-friendly and economically viable solution for solid waste management that will also help in organic farming and sustainable development. In rural areas the paper waste can be spiked with dung for successful vermicomposting and in urban and civilized areas food and kitchen waste can be taken for this purpose. There are several benefits of this model of paper waste vermicomposting:

i) Waste processing is simple, less labour intensive, economically and environmentally feasible.

ii) Very small to huge quantities of paper waste can be recycled.

iii) It can be operated anywhere in small or bigger space at houses, gardens, institutions, nurseries, temples, hostels in cities, towns and villages.

iv) Different types of waste paper and cardboard can be handled. Even dirty and unclean paper (paper smeared with food waste) can be used. The plastic lamination gets separated and can be removed later.

v) Paper waste can be mixed with wide range of other waste additives, dung, food and kitchen waste, temple (floral), sewage sludge, market (vegetable and fruits) waste etc.

vi) Like other vermicompost, paper waste vermicompost has favourable NPK values organics (certain enzymes and hormones) secreted by the worms in their castings. These compounds are believed to stimulate plant growth and suppress pathogens (Ismail, 1997; Abbasi and Ramasamy, 2001).

## Conclusion

Adverse effects of chemical farming have started their appearance on the human health. Numerous diseases are developing at a fast pace out of which several are non-curable. We all are taking residues of pesticides and traces of heavy metals in the food and vegetables. The outer shining of the fruits and grains has been increased but these have lost the inner nutritional status. All are toxicated. Use of organic manure, green manure and compost could only safe our life and could provide a way for sustainable agriculture for a sustainable development in the years to come. It can be accelerated by giving training and awareness programs at village and block levels and even at the doors of the farmers. Its benefits do not fade years after years; it has no adverse effects on human beings and the environment. Replacement of chemical fertilizers with organic composts and vermicompost is the major tool for sustainable development. Bioremediation of waste biomass through earthworms (vermicomposting) can also resolve the problems of waste management and can help in environmental conservation. Practice of organic farming should be promoted for sustainable development. Vermicomposting should be promoted for this purpose and all sorts of organic wastes including paper waste should be included as raw material for vermicomposting.

## Acknowledgements

We are thankful to the Head, School of Studies in Zoology, Jiwaji University, Gwalior, for providing all necessary facilities; UGC, New Delhi, for the sanction of SAP and meritorious fellowship and DST for the "FIST" programme.

## References

Abbasi,S.A. and Ramasamy, E.V. (2001): Solid waste management with earthworms. *Discovery Publishing House,* New Delhi, 178.

Adney, W. S., Rivard, J. C., Shing, M. and Himmel, M. E. (1991): Anaerobic digestion of lignocellulosic biomass and wastes. Cellulases and related enzymes. *Appl. Biochem. Biotechnol.,* **30:** 165-183.

Agrawal, O.P. (2005a): Vermicomposting practice-conversion of garbage into gold. *STTPP WAMR,* pp188-196.

Agrawal, O.P. (2005b): Vermitechnology for all-round sustainable development. Lead Paper Presented in National Seminar on Composting & Vermicomposting held at CSRTI, Mysore, October 26-27, pp 39-47.

Agrawal, O.P. (2009): Significance of Earthworms in Past, Present and Future Perspectives. In: Earthworm Ecology and Environment. (Ed. S.M. Singh), pp 51-73. IBDC., Lucknow.

Bhawalkar, U.S. (1993): *Turning Garbage into Gold***.** *BERI,* Pune. 40 pp.

Basheer, M. & Agrawal, O.P. (2013): Management of paper waste by vermicomposting using epigeic earthworm, *Eudrilus eugeniae* in Gwalior, India. *Int. Curr. Microbiol. App. Sci.* **2** (4): 42-47.

Busswell, J. A. and Chang, S.T. (1994): Biomass and extracellular hydrolytic enzyme production of six mushroom species grown on soyabean waste. *Biotechnology letters,* **16:**1317-1322.

Butt, K. R. (1993): Utilization of solid paper mill sludge and spent brewery yeast as a food for soil-dwelling earthworms. *Bioresource Tech.* **44:** 105-107.

Butt, K. R., Frederickson, J. and Morris, R. M. (1995): An earthworm cultivation and soil inoculation technique for land restoration. *Ecological Engineering,* **4:** 1-9.

Ceccanti, B. and Masciandaro, G. (1999): Vermicomposting of municipal and papermill sludges. *Biocycle,* **6:** 71±72.

Dandotiya, P. and Agrawal, O. P. (2013): Vermicomposting of food and household organic waste using epigeic earthworm *(Eudrilus eugeniae). International J. Curr. Res.* **5(10):** 3016-3019.

Dandotiya, P. and Agrawal, O. P. (2014): Stabilization of Vegetable Market Waste through Vermicomposting. *International Journal of Science and Research,* **3(6):** 50-55.

Deolalikar A.V., Mitra A., Bhatlacharyee, S. and Chakraborty, S. (2005): Effect of vermicomposting process on metal content of paper mill solid waste. *J. Enviro. Sci. Eng.,* **47(2):**81-84.

Edwards, C.A. (1988): Breakdown of animal, vegetable and industrial organic wastes by earthworm. *Agric. Ecosyst. Environ.* **24:** 21-31.

Elvira, C., Dominguez, J., Sampedro, L. and Mato, S. (1995): Vermicomposting from the paper pulp industry. *Biocycle,* **3**: 62-63.

Elvira, C., L. Sampedro, E, Benitez and Nogales, R. (1998): Vermicomposting of sludges from paper mill and dairy industries with *Eisenia andrei*: a pilot scale study. *Biores. Technol.* **63:**205-211.

Elvira, C., Goicoechea, M., Sampedro, L., Mato, S. and Nogales, R. (1996a): Bioconversion of solid paper pulp mill sludge by earthworms. *Biores. Technol.,* **57:** 173-177.

Elvira, C., Sampedro, L., Dominguez, J. and Mato, S. (1996b): Vermicomposting of wastewater sludge from paper pulp industry with nitrogen rich materials. *Soil Biol. Biochem.,* **29:** 759-762.

Gajalakshmi, S., Ramasamy, E.V. and Abbasi, S.A. (2001a): Screening of four species of detritivorous (humus–former) earthworms for sustainable vermicomposting of paper waste *Environmen. Technol.* **22:** 679–685.

Gajalakshmi, S., Ramasamy, E.V. and Abbasi, S.A. (2001b): Towards maximizing output from vermireactors fed with cow-dung spiked paper waste. *Bioresour. Technol.* **79:** 67–72.

Gajalakshmi, S., Ramasamy, E.V. and Abbasi, S.A. (2002): Vermicomposting of different forms of water hyacinth by the earthworm *Eudrilus eugeniae* Kinberg. *Bioresour. Technol.* **82:** 165–169.

Gajalakshmi, S. and Abbasi, S.A. (2003): High-rate vermicomposting system for recycling paper waste. *Indian Journal of Biotechnology,* **2(4)** October: 613-615.

Gajalakshmi, S. and Abbasi, S.A. (2004): Neem leaves as a source of fertilizer-cum-pesticide vermicompost. *Bioresource Technol.* **92**: 291-296.

Ghatnekar, S.D., Kavian, M.F., Ghatnekar, G.S. and Ghatnekar, M.S. (1998): Managemant of solid wastes through vermiculture biotechnology. **In:** *Ecotechnology for pollution control and environment management. (*Ed. R.K.Trivadi and Arvind Kumar). *Enviro. Media. Karad,* pp. 59-67.

Gohil, M.B. (1995): Treatment of pulp and paper mill waste. Case studies in population management in industries. (Ed. *Trivedy R.K.) Environment Publications, Karsad,* pp. 16-20.

Gratelly, P., Benitez, E., Elvira, C., Polo, A. and Nogales, R. (1996): Stabilization of sludges from a dairy processing plant using vermicomposting. In: *Rodriguez-Barrueco, C. (Ed.), Fertilizers and Environment,* Kluwer, The Netherlands. pp. 341-343.

Ismail, S.A. (1997): *Vermicology:* The biology of earthworms Orient Longman, Ltd., Chennai, India.

Kale, R.D. (1986): Earthworm feed for poultry and aquaculture. **In:** *Organic Waste Utilization and Vermicomposting* ( Ed. Dash, M.C.; Senapati, B.K. and Mishra, P.C) Sambalpur University, India.

Kaur, A., Singh, J., Vig, A. P., Dhaliwal, S. S. and Rup, P. J. (2010): Composting with and without *Eisenia fetida* for conversion of toxic paper mill sludge to a soil conditioner. *Bio. Res. Tech.,* **101 (21):** 8192- 8198.

Milala, M.A., Shehu. B. B., Zanna, H. and Omosioda, V.O. (2009): Degradation of Agro waste by cellulose from Aspergillus candidus. *Asian J. Biotechnol.,* **1:** 51-56.

Masciandaro, G., Ceccanti, B. and Garcia, C. (2000): In situ "vermicomposting of biological sludges and impacts on soil quality". *Soil Biol. Biochem.*, **32:** 1015-1024.

Mupondi, L.T., Mnkeni, P.N.S. and Muchaonyerwa, P. (2010): Effectiveness of combined thermophilic composting and vermicomposting on biodegradation and sanitization of mixtures of dairy manure and waste paper. *African J. Biotech.*, **9(30):** 4754-4763.

Mupondi, L.T., Mnkeni, P.N.S. and Muchaonyerwa, P. (2011): Effects of a pre-composting step on the vermicomposting of dairy manure-waste paper mixtures. *Waste Management & Research*, **29(2):** 219-228.

Nagavallemma, K.P. Wani, S.P., Stephane. L., Padmaja, V.V., Vineela, C., Babu Rao, M. and Sahrawat, K.L. (2004): Vermicomposting: Recycling wastes into valuable organic fertilizer. Global Theme on Agre-ecosystems. Report no. 8. Patancheru 502 324, Andhra Pradesh, India: International Crops Research Institute for the Semi-Arid Tropics. p-20 .

Nath, Gopabandhu and Deb U.K. (2008): Vermicomposting- an effective technique for paper mill solid waste utilization and value addition. *IPPTA*, **20(2):** 127-132.

Parray, M. A. (2012): Stimulatory influence of some additives on vermicomposting. M. Phil. Dissertation. Jiwaji University, Gwalior, Madhya Pradesh, India.

Pramanik, P and chung, R.Y. (2011): Changes in fungal population of fly ash and vinasse mixture during vermicomposting by *Eudrilus eugeniae* and *Eisenia fetida*: Documentation of cellulase isozymes in vermicompost. *Waste Management*, **31(6):** 1169-1175.

Rasal, P.H., Kalbhor, H.B., Shingte, V.V., Patil, P.L. (1988): Development of technology for rapid composting and enrichment. Biofertilizers – Potentialities Problems, (Ed. Sen, S.P. and P. Palit). *Plant Physiology Forum and Naya Prakash*, Calcutta, India, 255-258.

Raghuveer, S. and Shastri, C.A. (1995): Studies on biological treatment of pulp and paper mill waste water. *Waste Treatment Plants. Narosa Public House, New Delhi*. p. 401-405.

Ravi Kumar, P., Jayaram, P. and Somashekhar, R.K. (2009): Assessment of the performance of different compost models to manage urban household organic solid wastes. *Clea Techn. Environ. Policy*. **11:** 473-483.

Satchell, J.E. and Martin, K. (1984): Phosphatase activity in earthworm faeces. *Soil Biol. Biochem.*, **16:** 191–194.

Sunitha, N.S. (2012): Larm as a Necessity during Composting and/or Vermicomposting of Kitchen Refuses. *Universal J. Environ. Res. Techn.*, **2(1):** 97-100.

Suriyanarayanan, S., Mailappa, A.S., Jayakumar, D., Nanthakumar, K., Karthikeyan, K. and Balasubramanian, S. (2010): Studies on the Characterization and

Possibilities of Reutilization of Solid Wastes from a Waste Paper Based Paper Industry. *Global J. Environ. Res.* **4 (1):** 18-22.

Senapati, B.K. (1992): Vermitechnology: An option for recycling of cellulosic waste in India. **In:** *New Trends in Biotechnology.* (Ed. M.S. Subba Rao, C. Balgoplan and S.V. Rama Krishnan), p. 347-358. Oxford and IBH Publishing Co. Pvt. Ltd.

Talashilkar, S.C. and Powar, A.G. (1998): Vermitechnology for eco-friendly disposal of wastes. Ecotechnology for pollution control and environment management. *Indian J. Environ. & Ecoplan.* **7(3)**:535-538.

Umamaheshwari, S., Balamurugan, V. and Vijayalakshmi, G.S. (2009): Composting of paper mill sludge using native earthworms and impact on the growth and yield parameters of black gram, **In:** *Earthworm Ecology and Environment. (Ed. S.M.Singh): IBDC Publishers, Lucknow:* 135-143.

Vasanthy, M., Karthika, A. and Seethadevi, G.(2011): Determination of the potentiality of *Eisenia fetida* to degrade disposable paper cup waste- a solution to solid waste pollution. Proc. Intern. Conf. on Solid Waste: Moving Towards Sustainable Resource Management, Hong Kong SAR, P.R. China. p. 186-198.

# 4

# Kitchen Waste - Routine Home Waste: A Good Source for Organic Manure

*P S Chaudhuri*

*Department of Zoology, Tripura Central University, Suryamaninagar, Tripura*

## Kitchen Waste

Household kitchen waste is one of the major sources of municipal solid wastes. In India domestic waste is mostly of organic nature and contributed 70%-80% of urban solid wastes (Kale,1998). In Tripura each household of four member family generates 0.5 kg of kitchen wastes per day. Kitchen waste being nitrogen rich, heterogeneous and very soft plant materials are easily decomposed by microorganisms under tropical climatic conditions. One common practice of the semi-urban and rural people is to throw out these wastes in nearby drain making the latter a breeding ground for the mosquitoes.

Vermicomposting is a method of conversion of organic wastes into organic manure through Synergistic actions of bacteria, fungi, micro arthropods and earthworms. As earthworms play major role in the process, it is called vermicomposting. Kitchen wastes vermicomposting seems to be advantageous over the process of conventional composting. This biological process reduces smell of the organic wastes and improves the physical structure of the product *i.e.* vermicompost. Besides these, mineralization and humification brought out by microorganisms are accelerated by the passage of organic materials through the gut of earthworms, probably in presence the symbiotic intestinal micro flora and enzymes present in the gut of earthworms. The differences between worm-worked and worm less wastes under laboratory conditions indicate that stabilization had proceeded faster in the presence than in absence of worms.

## Process of kitchen waste vermicomposting

Essential items of kitchen waste vermicomposting are epigeic earthworms (*Perionyx excavatus*), large earthen bowls or cemented rings, kitchen wastes, shaded place and water source. Chief ingredients of kitchen wastes were vegetable skin (potato, onion), leaves of cabbage, parts of cauliflower and carrot, fruit skin (banana), egg shells and drained liquids of boiled rice.

Locally available epigeic earthworms, *Perionyx excavatus* (Indian blue) were preferred to other composter species like *Eisenia fetida* and *Eudrilus eugeniae* because the former species are readily available in kitchen wastes or cow dung deposit sites under shaded conditions in nature. These earthworms are bluish red species (100-180 mm length) occurring in clumps in the cow dung waste of Tripura.

Kitchen wastes were mixed properly and subjected to decomposition for ten days without earthworms. Pre-composting was carried out because *Perionyx* did not prefer fresh wastes. After ten days of pre-composting, 10 kg wastes were loaded in the earthen bowls where 600 worms (300 g) were introduced .Water was sprayed every five days to keep 50 % moisture in the substrate. Kitchen wastes contained too much water, so regular watering was not needed. Feeding-cum-casting activities started soon after introduction of the worms in the substrate. The upper layer of organic wastes was turned gently every ten days of interval. Foul odor disappeared ten days after vermicomposting process started. With the lapse of thirty days the refuse was changed into soft, spongy, sweat smelling dark brown compost. Fungal mycelia and large number of collembolan and oribatid mites appeared during decomposition of kitchen wastes in presence of earthworms. Nutrient content of kitchen waste vermicompost is given in Table - 1.

**Table 1: Nutrient contents of kitchen waste vermicompost**

**(Three replications)**

| Parameters | Kitchen waste (0 day) | Vermicompost (30 days) |
|---|---|---|
| pH | 10.0 ± 0.003 | 7.59 ± 0.15 |
| Organic carbon (%) | 36.8 ± 0.71 | 10.48 ± 0.32 |
| Total N (%) | 3.49 ± 0.06 | 1.67 ± 0.13 |
| Total P (%) | 0.89 ± 0.02 | 1.09 ± 0.02 |
| Total K (%) | 2.18 ± 0.02 | 0.85 ± 0.02 |
| Total Ca (%) | 4.73 ± 0.26 | 2.83 ± 0.16 |
| Total Mg (%) | 0.54 ± 0.01 | 0.40 ± 0.01 |

**Remarks**: Earthworms may selectively ingest high quality residues but their relative effects on decomposition may be greater on low quality residues which earthworms fragment, inoculate with microbes and incorporate in soils. High quality substrates like kitchen wastes tend to decompose rapidly even in absence of earthworms. It is felt that greatest gains in containment of nutrients can be achieved by placement of bulky carbonaceous wastes (eg. straw, saw dusts, rice bran etc.) to increase the C: N ratio and extend the biological activity, maximizing earthworm yields, thus proportion of nutrients extracted (Bhattacharjee 2002).

## Application of Kitchen Waste Vermicompost on Up Land Rice Productivity in Tripura

A recent field trial was conducted during Kharif season at ICAR research complex, Lembuchhera, West Tripura, on up land paddy (var.TRC 87-251) with various doses of vermicompost prepared from Kitchen waste-cow dung mixture (1:4 W/W) using *Perionyx excavatus* earthworms .The field experiment was laid out in RBD technique comprised of 12 treatments, each treatment having four replica. The vermicompost (VC) applied to the field had organic carbon 7.22%, nitrogen 1.88 %, C/N ratio 3.84, phosphorus 0.20 % and potassium 0.38 %. The NPK dose was 60 kg N, 30 kg P and 30 kg K per hectare in the form of urea, single super phosphate and muriate of potash, respectively. The treatments as applied in the experiment were $T_0$ (control), $T_1$ (2.5 tons VC/ ha), $T_2$ ($T_1$ + NPK), $T_3$ (5 tons VC/ ha), $T_4$ ($T_3$ + NPK), $T_5$ (10 tons VC/ ha), $T_6$ ($T_5$ + NPK), T 7 (15 tons VC/ ha), $T_8$ ($T_7$ + NPK), $T_9$ (20 tons VC/ ha), $T_{10}$ ($T_9$ + NPK), $T_{11}$ (NPK).

Vermicompost had a positive effect on the fertility status of soils, changing its physico-chemical properties and thereby leading to improved crop yield. From the field trial, the highest level of straw yield (56.5 % increment) was observed following application of 10 tons vermicompost in combination with NPK. Kale and Bano (1986) reported that vegetative growth of paddy (IR-20) was influenced by the vermicompost (Vee comp. E 83 UAS) in a better way than the chemical fertilizer.

Application of 15 tons vermicompost per hectare produced the highest (81.7 %) increase in yield of rice as compared to control. The effects of 10 to 15 tons vermicompost per hectare and supplementation of NPK with 5 to 10 tons of vermicompost per hectare on the increase in grains and straw yield were not significantly different. Vasanthi and Kumaraswamy (1999) have also reported that vermicompost at @ 5 t/ha would be sufficient for the production of rice crop when applied with recommended levels of N,P and K. Yield of paddy was comparatively much higher in the plots treated with N , P , K alone (**Table - 2**). According to Brady (1994) nitrogen encourages vegetative growth and is a regulator that governs to considerable degree the uptake of phosphorus and potassium, which are important for seed production. Kale *et al.*, (1992) and Zhao Shi-Wei and Huong Fuzhen (1992) reported that chemical fertilizer application along with vermicompost in cereals promotes nutrient uptake for better yield. Significant increase in the uptake of nutrients following supplementation of NPK in 5 to 10 tons vermicompost ($T_4$ and $T_6$) compared with vermicompost alone in $T_5$ and $T_7$ was recorded that corroborates with the findings of Kale *et al.*, (1992) and Zhao Shi- Wei and Huong Fuzhen (1992) who have reported that the application of chemical fertilizers along with vermicompost promotes nutrient uptake in cereals for better yield. This indicates that vermicompost reduces the loss of nutrients through leaching from soil by changing the physico-chemical properties of soils (Bhattacharjee, 2002). Aggregation of soil and its water use efficiency improved with increasing doses of vermicompost up to 10 to 15 tons vermicompost per hectare. With increase in soil aggregation, the number of micropores and thereby aerobicity and water holding capacity of the soil probably be increased, thus providing the characteristics of a healthy soil, which ultimately increased the water use efficiency of the plants. This suggests that effect of vermicompost on crop production is not solely nutritional but also to provide ideal physical structure of the soil to increase crop yield.

**Table - 2: Biomass and grain yield of upland rice (var TRC-87-251) and water use efficiency following application of different doses of vermicompost**

(Bhattacharjee 2002)

| Treatments | Dry matter (g plant$^{-1}$) | | Grain yield quintal ha$^{-1}$ | Straw yield quintal ha$^{-1}$ | Water use efficiency Kg ha-mm$^{-1}$ |
|---|---|---|---|---|---|
| | Tillering | Flowering | | | |
| $T_0$ (Control) | 1.39 | 4.09 | 23.39 | 40.32 | 6.79 |
| $T_1$ (2.5t VC ha$^{-1}$) | 1.78 | 4.85 | 29.24 | 34.77 | 8.70 |
| $T_2$ ($T_1$ + NPK) | 1.79 | 3.98 | 28.31 | 52.38 | 7.92 |
| $T_3$ (5t VC ha$^{-1}$) | 1.85 | 3.36 | 31.03 | 40.80 | 9.04 |
| $T_4$ ($T_3$ + NPK) | 2.81 | 4.24 | 37.95 | 57.74 | 10.90 |
| $T_5$ (10t VC ha$^{-1}$) | 1.54 | 4.90 | 39.61 | 59.76 | 12.65 |
| $T_6$ ($T_5$ + NPK) | 3.43 | 5.26 | 39.62 | 63.11 | 11.55 |
| $T_7$ (15t VC ha$^{-1}$) | 1.85 | 5.13 | 42.49 | 59.60 | 12.72 |
| $T_8$ ($T_7$ + NPK) | 2.15 | 4.26 | 37.28 | 53.48 | 10.56 |
| $T_9$ (20t VC ha$^{-1}$) | 1.88 | 4.96 | 34.52 | 47.16 | 9.40 |
| $T_{10}$ (T9 + NPK) | 3.15 | 6.30 | 35.69 | 60.60 | 9.83 |
| $T_{11}$ (NPK) | 1.46 | 5.76 | 25.87 | 50.04 | 6.89 |
| Mean | 2.09 | 4.75 | 33.75 | 51.64 | 9.74 |
| CV (%) | 30.8 | 16.5 | 17.2 | 17.1 | 19.76 |
| CD (5%) | 1.46 | 1.67 | 8.74 | 14.61 | - |

*VC – Vermicompost*

## References

Bhattacharjee, G. (2002): Earthworm resources and waste management through vermicomposting in Tripura, *Ph.D. Thesis, Tripura University*

Brady, N.C. (1984): The nature and properties of soils. *Euresia publishing House (P) Ltd. New Delhi.*

Kale, R.D. (1998): Earthworms: Cinderella of organic farming *Prism Books Pvt. Ltd. Bangalore.*

Kale, R.D. and Bano, K. (1986): Field trials with vermicompost (Vee comp E.83 UAS) an organic fertilizer. In: Dash, M.C., Senapati, B.K. and Mishra,

P.C. (ed.) *Proc. Nat. Sem. Org. Waste Utilz. Vermicomp. Part B: Verms and Vermicomposting, Five Star printing press, Burla, Orissa 151-156.*

Kale, R.D., Mallesh, B.C., Bano, K. and Bagyaraj, D.J. (1992): Influence of vermicomposting application on the available macronutrients and selected microbial population in a paddy field. *Soil Biol. Biochem.* 24: 1317- 1320.

Vasanthi, D. and Kumaraswami, K. (1999): Efficiency of vermicompost to improve soil fertility and rice yield. *J. Indian Soc. Sci.* **47(2)** 268-272.

Zhao, Shi-Wei and Huong Fu-Zhem (1992): The nitrogen uptake efficiency from N15 labeled chemical fertilizer in presence of earthworm manure (Cast) 539-542 In: Veeresh, G.K., Rajagopal and Viraktamath CA (eds.). Advances in management and conservation of soil fauna, *Oxford & IBH, N. Delhi.*

# 5

# Pressmud: A Rich Source of Organic Manure

*K. Parthasarathi*[1] *and L.S. Ranganathan*[2]

[1]*Department of Zoology, Annamalai University, Annamalai Nagar- 608 002, Tamilnadu (India)*

[2]*Former Professor and Head of the Department of Zoology & Dean, Faculty of Science*

India, the second most populous country in the world, depends predominantly on agriculture. Despite of the huge population and poverty ruling the masses, the country is managing to feed most of them, because of its vastness and richness of the soil. The Indo-Gangetic delta is one of the most fertile spots on this planet but industrialization, urbanization and pollution have reduced the fertility of the soil.

In India, the area of productive agricultural land is about 143 million hectares, of which 83 million hectares suffer from various degrees of soil degradation. Hence conservation of soil health, fertility, plant productivity and wise use of soil are very essential today. It has been proved that indiscriminate use of inorganic fertilizers results in decreasing the soil fertility and increasing the soil acidity with depletion of organic humus content in addition to poor crop quality.

Organic farming is a system where readily soluble, synthetic fertilizers and biocides are avoided during cultivation. It relies on crop rotation, crop residues, animal manure particularly vermicompost, legumes, green manures, bioremediation and biological pest control to maintain productivity and tilth. Use of organic manures to meet the nutrient requirements of crop would be an inevitable practice in the years to come since organic manures not only improve the soil's physical, chemical and biological properties but also improves the moisture holding capacity of soil, thus resulting in enhanced crop productivity along with better quality of crop produce.

Wastes become wealth if properly managed. Organic waste is anything that comes from plants or animals that is biodegradable. All kinds of organic wastes are the raw materials for organic manure production. To derive beneficial product from wastes, numerous techniques are available and all the techniques are mainly based on the concept of recycle, reuse and recovery of resources. Vermicomposting, a novel technique of converting decomposable organic waste into valuable vermicompost through earthworm activity is a faster and better process when compared with the conventional methods of composting. Vermicomposting of organic wastes using earthworm living on the surface soil (epigeic) is one of the recent techniques for the recycling of organic wastes and it is a viable, eco-friendly, efficient, ecologically sound method for waste management and manure production, and also an exceedingly benign technology which can very substantially contribute to global warming control *i.e.* by means of preventing methane formation (methane has 25 times greater global warming potential than $CO_2$) and returns the carbon that was fixed in phytomass to the soil in a way that enhances the soil's productivity.

## Pressmud – A Rich Resource for Vermi-Manure Production

Since 19$^{th}$ century, sugarcane is grown in various countries primarily for sugar production. India has the largest area under sugarcane cultivation amongst the cane growing countries of the world. Sugar industry in India is one of the oldest industries and has great relevance in the context of rural development. Sugarcane is one of the important cash crops in India and occupies about 3.3 million hectares representing two per cent of the cropped area. More than 45 million of sugarcane growers in the India and about 65% of the rural population depend on this agro-based industry. Sugar industry is the largest amongst the processing industries next only to textile industry and produces about 14.5 million tons of sugar in recent years. Sugarcane occupies a prominent place by providing an income of about 22,000 crore per annum to the cane growers and about Rs. 220/- crore and Rs. 100/-crore per annum as taxes for Union and State Governments respectively. As regards Tamilnadu, the average cane yield is about 103.0 t/ha and this crop is in cultivation in about 2.60 lakh hectares supplying canes to 35 sugar factories. In the last decade, the sugar recovery had increased from 8.5 to 9.5 %.

Sugar industries generate large amount of waste by-products in the form of pressmud, bagasse, molasses and spent wash per day. Pressmud (filter-cake), a residue left over from sugarcane after extraction of juice, is a soft, spongy, light weight, amorphous, dark brown to black material. Depending on the quality of cane and the extraneous matter that come with sugarcane, pressmud is about 3.5 – 4.5% of total cane crushed. The annual production of pressmud, at present, is more than 10.0 million tones (Figs. 1 & 2). The raw sugarcane juice obtained during cane milling for sugar manufacture contains suspended and dissolved impurities like fibers, wax, oil, resin and ash. Two types of processes such as 'sulphitation' and 'carbonation' are commonly used to purify the cane juice. The clarification of cane juice in the sulphitation process is carried out with the help of lime and $SO_2$ and in the carbonation process with lime and $CO_2$. The production of sulphitation and carbonation pressmud is about 4% and 7% of the quantity of cane crushed in a sugar factory following sulphitation or carbonation processes respectively. The chemical composition of pressmud varies with different industries with varying factors such as type of soil, cane variety, type of harvest, level of juice extraction, amount of lime and other products used in clarification, pH, temperature, fertilizer and nutrients availability. The composition of pressmud generated from the sugar factory is given in Table -1.

**Table- 1: Composition of Pressmud (in %)**

| | |
|---|---|
| Moisture | 50-60 |
| Crude wax | 5 – 15 |
| Fiber | 15 – 30 |
| Protein | 3 – 15 |
| Sugar | 4 – 16 |
| Nitrogen | 1.2 – 3 |
| Phosphorus | 0.60 – 3.5 |
| Potassium | 0.30 – 1.8 |
| Total ash | 9 – 12 |

Though the press-mud has good nutrient value to be used as manure farmers were reluctant and afraid to apply this due to the fear of crust formation on the soil and bringing about pH variations. Vermicomposting of pressmud makes it possible to be used for maintaining sustainable soil health, fertility, waste degradation, soil reclamation, land restoration practices and gainful utilization of the by-product generated from sugar factory. The sugar factories have already invested hugely for production of manure, the process of which is duly approved by the Central Pollution Control Board (CPCB). In case, such a thing is allowed to happen, the mills would not be able to produce the soil friendly manure and all the useful nutrients would be burnt or lost.

Keeping in view the benefits of the manure and good availability of the same, it would be of interest to the farmer's community that bio-composting or vermicomposting press-mud is encouraged.

## Production of Pressmud Vermi-Manure

### i) Preparation of pressmud for vermicomposting

Pressmud when formed in the sugar mill is in the form of slurry, hot and with bad odour. It is collected from the sugar factories and transported to the place where it is processed for its transformation into vermi-manure (Figs. 1& 2). The first step is to remove the odour by exposing it to sunlight in a open place. One month old pressmud is obtained from the sugar mill. It is again sun-dried for few days and cured by sprinkling water, repeatedly, to remove odour and heat. It is maintained at 60-70% moisture and 75% relative humidity (Fig. 3).

**Fig. 1 & 2: Generation and transportation of pressmud from sugar factory**

**Fig. 3:** Curring of pressumed for vermimanure production

### ii) Vermicomposting of pressmud on small scale

About 75-100 worms (300 g) of sexually immature, pre-clitellate (25 –30 days old) *Eudrilus eugeniae* (Kinberg) were inoculated in cement tank (sized 50 x 35 x 30 cm), containing 5 kg cured, dried and pre composed (for 3 days) pressmud (Fig. 4). Any indigenous, surface living, litter eating worm could be used. *Lampito mauritii, Eisenia fetida* and *Perionyx excavatus* are good composting worms. Earthworm's growth (biomass), reproduction (cocoon and hatchling number) and vermicompost recovery rate were recorded for $60^{th}$ days. The cocoon, hatchlings and vermicompost were hand sorted and counted. The formed vermimanure was collected periodically and used for further studies like analysis of nutritional status and its vermimanure efficacy. The nutritional status of the pressmud vermimanure is given in Table - 2 and it should be noted that vermicomposting had enhanced the available carbon, nitrogen and potassium and important micronutrients such as iron, manganese and zinc. The essence of vermicomposting pressmud is to make available macronutrients and micronutrients that could be easily taken by the plants. During this process, beneficial microbes (to the plants) were also enhanced contributing to better soil fertility. Also the pressmud contributes to the worm better biomass, cocoon production, hatching rate and more vermimanure than cow dung (Fig. 5). Hence pressmud functions as a good substrate for vermiculture and bulking material for vermicomposting and seems to be better than cow dung.

**Table 2: Chemical characterization of pressmud vermi-manure of *E. eugeniae***

| Parameter | Pressmud | Pressmud vermi-manure | % changes |
|---|---|---|---|
| pH | 7.80 ± 0.04 | 7.52 ± 0.07 | -3.59 |
| EC (mnoh /cm) | 0.36 ± 0.006 | 34.0 ± 0.01 | -5.50 |
| OM % | 53.40 ± 0.03 | 33.38 ± 0.05 | -37 |
| OC % | 31.33 ± 0.06 | 26.57 ± 0.03 | +50.8 |
| N % | 1.20 ± 0.0001 | 1.81 ± 0.003 | +64 |
| P % | 3.44 ± 0.007 | 5.071 ± 0.005 | +32 |
| K % | 0.62 ± 0.02 | 0.76 ± 0.03 | +27 |
| Ca % | 2.56 ± 0.005 | 4.87 ± 0.002 | +47 |
| Mg % | 0.55 ± 0.0003 | 1.42 ± 0.06 | +61 |
| S % | 0.66 ± 0.0003 | 0.26 ± 0.0009 | -60 |
| Na % | 0.26 ± 0.0003 | 0.34 ± 0.0001 | +23.5 |
| Fe (ppm) | 184 ± 0.05 | 191 ± 42.57 | +8.15 |
| Mn (ppm) | 21.0 ± 0.03 | 32 ± 4.05 | +52.4 |
| Zn (ppm) | 61 ± 0.07 | 62 ± 131 | +1.64 |
| Cu (ppm) | 17 ± 0.0006 | 15 ± 0.005 | -11.76 |
| C:N Ratio | 26:1 | 14:1 | -70 |
| C:P Ratio | 9:1 | 5:1 | -42 |
| CRP | 7.5 ± 0.13 | 11.3 ± 0.10 | +50.8 |

*% increase (+)/decrease (-) over pressmud*

**Pressmud in cement tank**

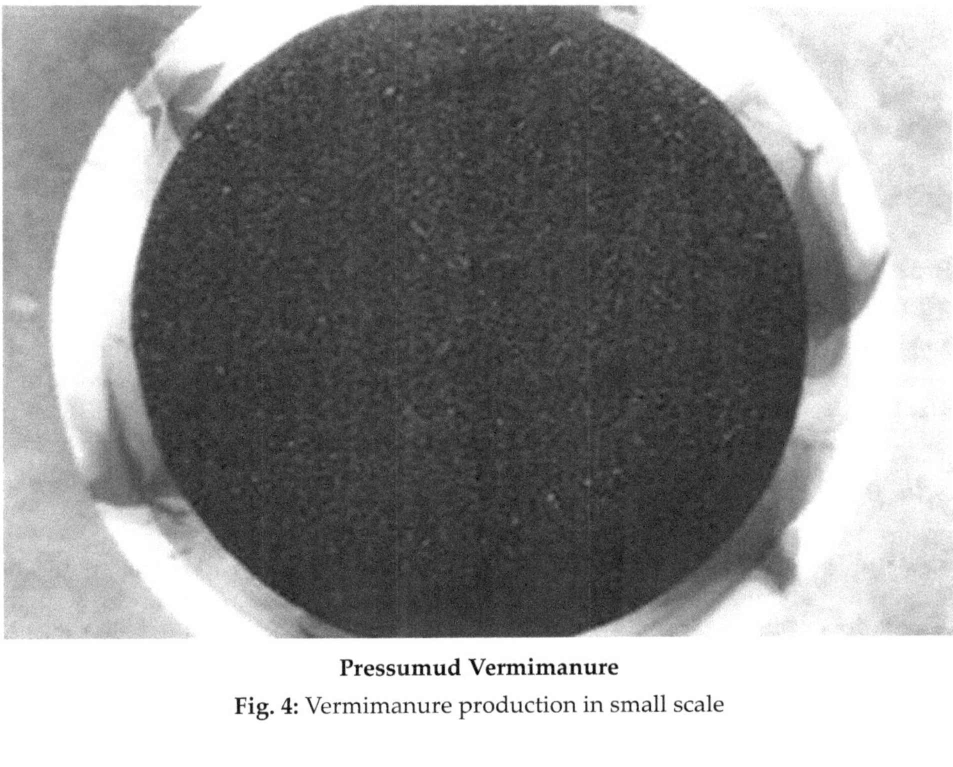

**Pressumud Vermimanure**

**Fig. 4:** Vermimanure production in small scale

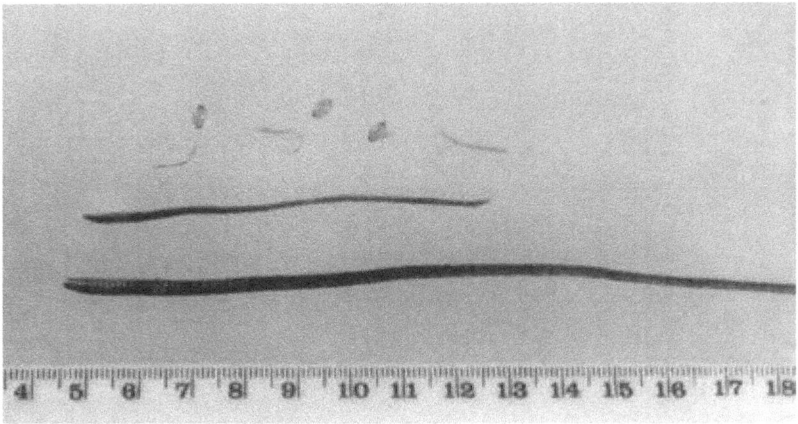


**Fig. 5:** Cocoon, hatchlings, adult worm of E. *eugeniae* in pressmud during vermicomposting

### iii) Vermicomposting of pressmud on large scale

Preparation of vermi-manure from pressmud on a large scale could easily be performed by making pits. These pits of 150 x 90 x 90 cm dimension were dug on raised ground to prevent inundation. The inner surface of the pit was lined by thick polythene sheet to prevent the escape of worms and entry of other soil organisms into the pit. The bottom of pit was filled with broken bricks (5 cm) and mother soil (3 cm). A thatched shed was provided as shelter against sunlight and rain (Fig. 6). Instead of pits, pressmud could also be composted in large wind-rows in open space but protected from sun, rain and predators. Care must be taken to protect earthworms from birds, lizards, ants, white ants, rat, bandicoot, shrews and snakes.

## Efficacy of pressmud vermi-manure on soil quality, growth, yield and quality of crops

Field experiments were conducted on clay loam soil at Sivapuri, Chidambaram, Cuddalore district, Tamil Nadu, India, to evaluate the efficacy of pressmud vermi-manure (VM) on the soil quality characteristics, and on the yield and quality characteristics of pulse – beans (*Phaseolus vulgaris*) and chilli – *Capsicum annuum* in comparison to inorganic fertilizers (NPK). Application of VM (5 tons/ha) and VM supplemented with recommended dose of inorganic fertilizer (RDIF) (20:80:40 kg.ha) and (120:60:30 kg.ha) (w/w) had significantly increased the pore space, water holding capacity, cation exchange capacity, organic carbon, available N, P, K, other micro-macro nutrients – Ca, Mg, Na, Fe, Mn, Zn, Cu and microbial population - activity and humic acid content, reduced particle and bulk density, pH, EC in the field soil. Yield and quality (protein and sugar for pulse) (vitamin A, Vitamin C and Capsaicin content for chilli) were enhanced in the soil (Figs. 7 & 8) (Table- 3). On the contrary, the application of NPK alone has resulted in reduced porosity, compaction of soil, reduced organic carbon, reduced microbial population, microbial activity and humic acid content. On the whole, the results obtained indicates that the combined application of VM and RDIF show promising results in the cultivation and better for obtaining the higher quality of the crops. Further the VM reduce 50% RDIF to the crops and also enhances soil quality, yield and quality characteristics of crops than 100% application of RDIF on crops.

**Table 3: Effect of pressmud vermi-manure and NPK on yield and quality characteristics of crops ($p<0.05$)**

| Parameters | $T_1$ | $T_2$ | $T_3$ | $T_4$ | CD value |
|---|---|---|---|---|---|
| **Chilli** | | | | | |
| Yield (kg.ha$^{-1}$) a | 1819 | 1946 (+6.5)$^x$ | 2267 (+19.8)$^x$(+14.2)$^y$ | 2412 (+24.6)$^x$(+19.3)$^y$(+6.0)$^z$ | 7.05 |
| b | 618 | 686 (+9.9)$^x$ | 785 (+21.3)$^x$(+12.6)$^y$ | 813 (+23.9)$^x$(+15.6)$^y$(+3.4)$^z$ | 4.39 |
| Vitamin A (mg/ g)a | 55.6 | 63.2 (+12.0)$^x$ | 76.8 (+27.6)$^x$(+17.7)$^y$ | 81.5 (+31.8)$^x$(+22.5)$^y$(+5.8)$^z$ | 8.94 |
| b | 17.9 | 20.4 (+12.3)$^x$ | 24.6 (+27.2)$^x$(+17.1)$^y$ | 26.3 (+31.9)$^x$(+22.4)$^y$(+6.5)$^z$ | 3.46 |
| Vitamin C(mg/ 100 g) a | 62.3 | 76.6 (+18.7)$^x$ | 88.5 (+29.6)$^x$(+13.5)$^y$ | 94.7 (+34.2)$^x$(+19.1)$^y$(+6.6)$^z$ | 3.54 |
| b | 14.5 | 17.4 (+16.7)$^x$ | 20.2 (+28.2)$^x$(+13.9)$^y$ | 21.3 (+39.9)$^x$(+18.3)$^y$(+5.2)$^z$ | 3.00 |
| Capsacinin (%) a | 2.69 | 3.10 (+13.2)$^x$ | 3.83 (+29.8)$^x$(+19.1)$^y$ | 4.14 (+14.5)$^x$(+25.1)$^y$(+7.5)$^z$ | 1.91 |
| b | 1.20 | 1.49 (+19.5)$^x$ | 1.72 (+30.2)$^x$(+13.4)$^y$ | 1.88 (+36.2)$^x$(+2.07)$^y$(+8.5)$^z$ | 0.0007 |
| **Beans** | | | | | |
| Yield (kg.ha$^{-1}$) | 6750 | 9625 (+29.9)$^x$ | 11259 (+40.0)$^x$(+14.5)$^y$ | 14375 (+53.0)$^x$(+33.0)$^y$(+21.6)$^z$ | 0.26 |
| Sugar (mg.g$^{-1}$ dry wt.) | 210 | 240 (+12.5)$^x$ | 262 (+19.8)$^x$(+8.4)$^y$ | 290 (+27.6)$^x$(+17.2)$^y$(+9.7)$^z$ | 0.21 |
| Protein (mg.g$^{-1}$ dry wt.) | 290 | 310 (+6.5)$^x$ | 321 (+9.7)$^x$(+3.4)$^y$ | 382 (+24.1)$^x$(+18.8)$^y$(+16.0)$^z$ | 0.32 |

*$T_1$ – Control(without application of inorganic fertilizer/vermimanure); $T_2$ – Recommended dose of NPK (120:60:30 and 20:80:40 kg ha$^{-1}$); $T_3$ – Recommended dose of vermimanure (5 tons ha$^{-1}$); $T_4$ – 50% vermifertilizer supplemented with 50% NPK (w/w); a–Fresh(green) fruit; b – Dry fruit; CD value – Critical difference value; x - The figures in parentheses (+/-) indicates the percent increase / decrease over $T_1$; y - The figures in parentheses (+/-) indicates the percent increase / decrease over $T_2$; z - The figures in parentheses (+/-) indicates the percent increase / decrease over $T_3$.*

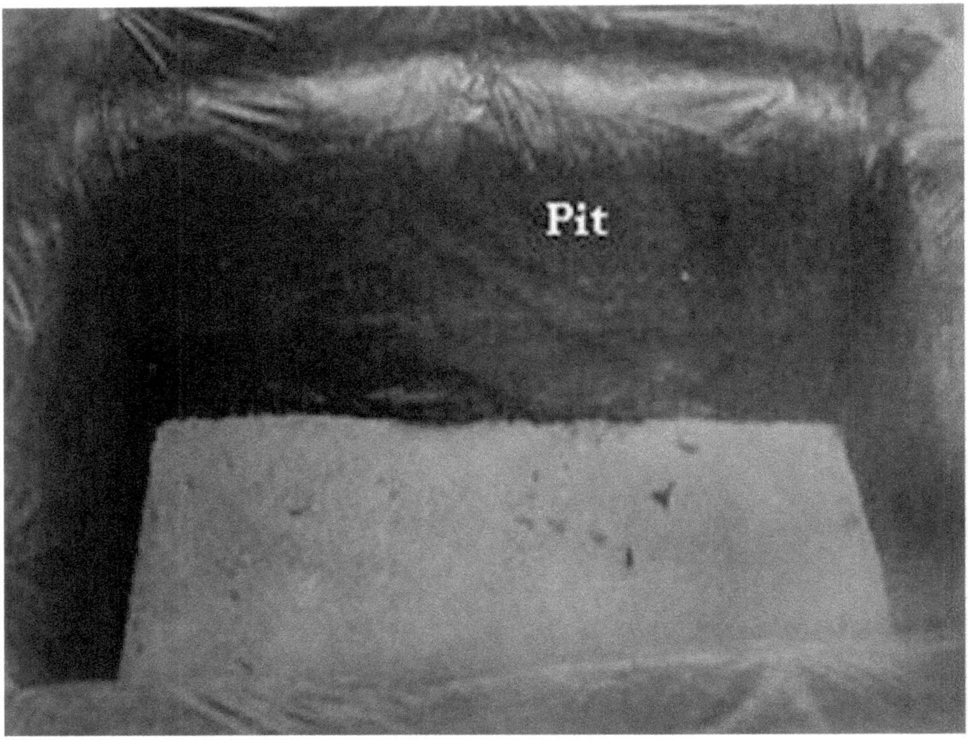
Pit

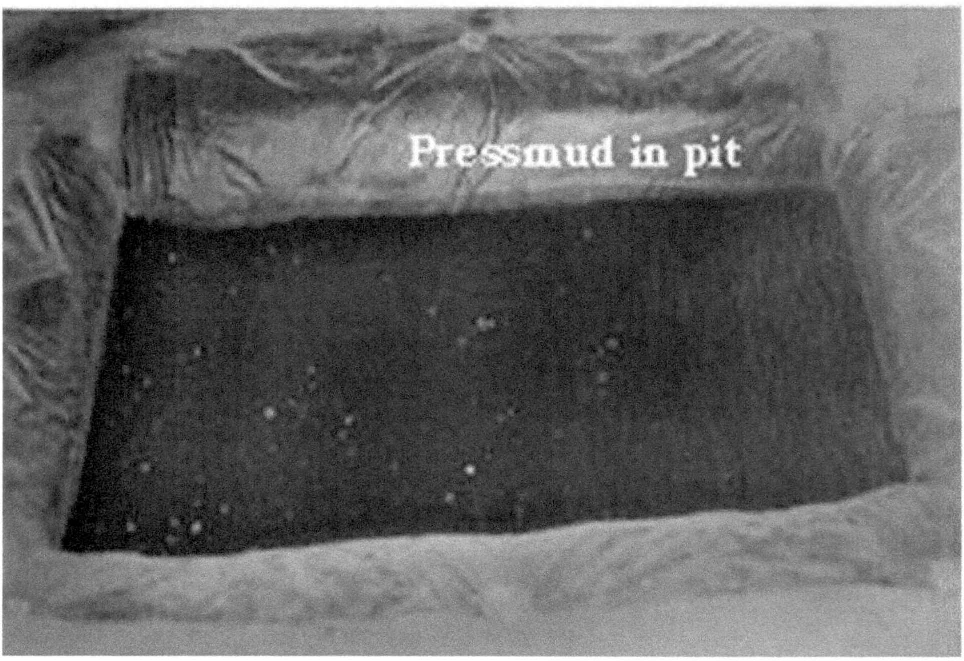
Pressmud in pit

**Fig. 6:** Vermimanur production in large scale

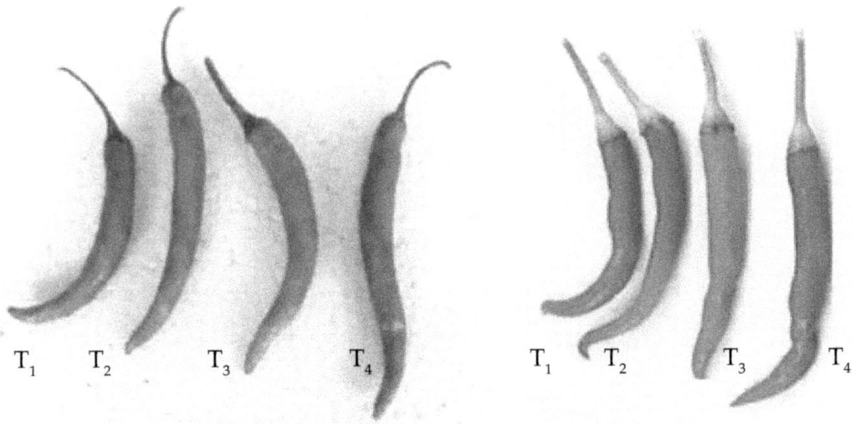


**Fig. 7:** Growth and yield characteristics of chilli

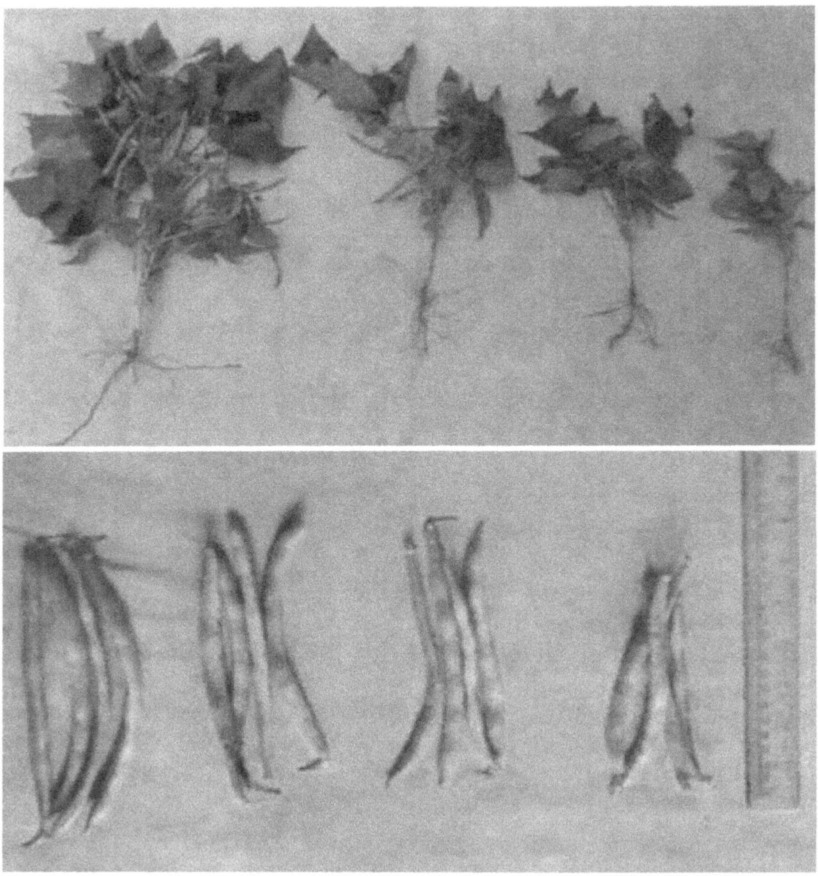

**Fig. 8:** Growth and yield characteristics of beans

## Conclusion and Recommendation

1. Dried, pre composted pressmud has to be given free to all the farmers attached to the sugar mill.
2. Supply of pressmud during off-season of crushing has to be ensured.
3. Initial quota of food composting local variety of worms have to be supplied free of cost, to prevent bogus intermediate persons.
4. A one day work-shop on vermicomposting could be conducted in sugar mills, colleges and Kisan Universities.
5. Buy-lack guarantee (for vermicompost) has to be extended.
6. Our experience has shown the sustainable maintenance of soil quality and productivity of various crops is much better with use of this pressmud vermi-manure. Therefore it is, in the interest of farmers and the agriculture sector to use pressmud as an alternative bulking material to cow dung for vermiculture and to produce vermi-manure.

## Acknowledgements

We thank the authorities of Annamalai University for providing facilities and financial assistance from the DST-SERB (SB/SO/AS-082/2013), New Delhi.

## References

Parthasarathi, K., M.Balamurugan, K.V.Prashija, L. Jayanthi and S.Ameer Basha (2016). Potential of Perionyx excavatus (Perrier) in lignocelllulosic solid waste management and quality vermifertilizer production for soil health. Int. J. Recycl. Org. Waste Agricult, (DOI 10, 1007/S 40093-016-0118-6).

Ranganathan,L.S.(2006).Vermibiotechnology-From Soil Health to Human Health. Agrobios,Jodhpur.

Parthasarathi, K. (2007). Influence of moisture on the activity of Perionyx excavatus (Perrier) and microbial – nutrient dynamics of pressmud vermicompost. Iran. J. Environ. Health. Sci. Eng., 4(3): 147-156.

K.Parthasarathi (2010) Earthworms – Life cycle, Compost and Therapy. Lap Lambert Academic Publishing AG & CO. KG. Germany (ISBN:978-3-8383-9575-3).

Parthasarathi, K. and L.S.Ranganathan (1999). Longivity of microbial and enzyme activity and their influence on NPK content in pressmud vermicasts. European Journal of Soil Biology, 35(3): 107-113.

Parthasarathi, K. and L.S.Ranganathan (2000). Aging effect on enzyme activities in pressmud vermicasts of Lampito mauritii (Kinberg) and Eudrilus eugeniae (Kinberg). Biology and Fertility of Soils, 30: 347-350.

Prashija, K.V., and Parthasarathi, K., (2016). Management of agroindustrial lignocellulosic wastes through vermitechnology and production of agronomic valid vermicompost. Int. J. Biotechnology Wellness industries, 5:153-167

# 6

# Fly Ash: A Rich Source for Organic Manure

*Gourab Roy, Wasim Iftikar*[1] *and G.N.Chattopadhyay*[2]

*Iswar Chandra Vidyasagar College, Belonia-799155, South Tripura, India*

[1]*CSISA Odisha Hub, ATIC Building, Directorate of Extension Education, OUAT, Bhubaneswar-751003, Odisha, India*

[2]*Ashirbad, Seba Pally, Santiniketan-731235, West Bengal, India*

## Fly Ash

Fly ash (FA) is the residue of combustion in different coal fired plants that enters the flue gas stream. With the consistently increasing numbers of coal-fired plants, large-scale generation of FA is presenting an acute waste disposal problem in different parts of the world (Bhattacharya and Chattopadhyay, 2001) of which India is an important component (Figure-1). Gradually deteriorating quality of coal due to extensive exploitation of mines is likely to aggravate this situation further, if clean coal technologies are not adopted widely. Generation of such huge amount of coal residues presents several problems, the most serious ones being environmental hazard and requirement of large areas for disposal of this waste material. Hence, the urgent and imperative needs to overcome these problems, not only through safe disposal but also through gainful recycling of these materials, cannot probably be over emphasized (Chattopadhyay and Bhattacharya, 2010). The seriousness of the problem has been recognized at global level and various possibilities of effective disposal and re-use of FA are being explored in different countries. Amongst these, application of FA for agricultural cultivation is a comparatively less attended in the farming land but has been showing promising results, especially for rejuvenating low fertile or wastelands (McCarty *et al.*, 1994; Schutter and Fuhrman, 1999; Kumar *et al.*, 2000). This benefit may be attributed primarily to alkaline nature of most of the FA, which may be used to improve the pH levels of acid soils, and also to its total content of different nutrient elements in varying concentrations (Roy *et al.*, 2014). However, the major problems associated with plant nutrition from FA are its inert nature owing to very poor microbiological activity in such material and the low occurrence of most of the nutrient elements in available forms (Chattopadhyay and Bhattacharya, 2010). Since vermicomposting assists in accelerating degradation of various kinds of waste through earthworms' gut microorganisms, the possibility of enhancing the status of availability of various plant nutrients in the fly ash for enhancing its acceptability in agricultural cultivation may be adopted at a wider level and should be initiated by the State and Central Governments too.

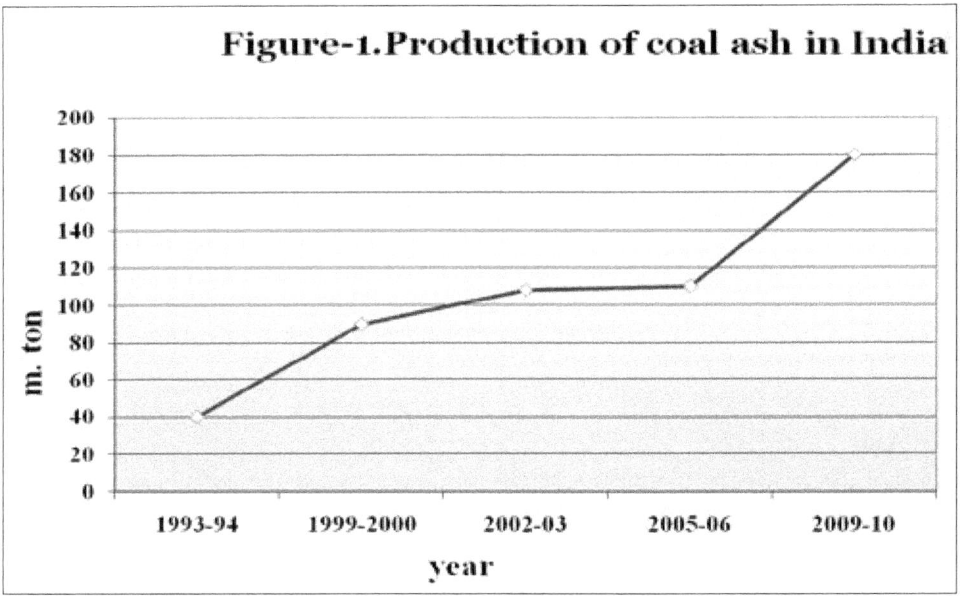


*Source: Bhattacharya and Chattopadhyay (2010).*

## Use of Fly Ash in Agriculture

Fly ash contains almost all the plant nutrients, excepting nitrogen and organic carbon, in considerable amount. Hence, it has been considered by many workers that this material may be used for several agriculture related activities as a source or carrier of plant nutrients (Page *et al.*, 1980; Tripathi, *et al.*, 1997). Considerable studies have been conducted in agriculture and forestry sectors to assess the effect of fly ash as a source of plant nutrients, which have shown that these ashes have some beneficial properties that could be utilized for agronomic purposes. Positive effects of FA application on the yield level of alfalfa, barley, different workers (Mulford and Martens, 1971; Hill and Lamp, 1980) have reported Bermuda grass, white clover etc. Total occurrence of many nutrients have also been reported to be comparatively higher than what are generally observed in arable soils (Table-1) and use of FA has been suggested for increasing the nutrient status of such soils (Kumar *et al.*, 2000).

In general, it has been observed that FA promotes better growth or yield of different crops up to a certain level of application of this material to soil and the beneficial effects tend to decline beyond that dose. These effects are, however, likely to vary from one crop to other and also in different soils. Use of coal ash on growth and nutritional status of spinach in two kinds of soils viz. black cotton and alluvial using various soil: ash mixtures has shown the shoot length, chlorophyll concentration and carbohydrate content of the plants to increase up to 25% of ash application (Kumari and Dubey, 2003). However, most of the studies carried out on the utilization of FA in agriculture deal with yields_of different crops due to application of various doses of ash and limited work have been carried out to provide scientific or technical backgrounds for such results.

**Table-1: General Characteristics of Indian Fly Ash and Arable Soils**

| Properties | Fly Ash | Soil |
|---|---|---|
| B.D.(g $cm^{-3}$) | < 1.0 | 1.33 |
| W.H.C. (%) | 35-40 | < 20 |
| Porosity (%) | 50-60 | < 25 |
| P (%) | 0.004-0.8 | 0.005-0.2 |
| K (%) | 0.19-3.0 | 0.04-3.0 |
| S (%) | 0.1-1.5 | 0.01-0.2 |
| Fe (mg $kg^{-1}$) | 36-133 | 10-300 |
| Zn (mg $kg^{-1}$) | 14-1000 | 2-100 |
| Cu (mg $kg^{-1}$) | 1-26 | 0.7-40 |
| Mn (mg $kg^{-1}$) | 100-3000 | 100-4000 |
| B (mg $kg^{-1}$) | 46-618 | 0.1-40 |

*Source : (Kumar et al., 2000)*

## Availability of Plant Nutrients in Fly Ash

In spite of these positive effects of fly ash application in agriculture, availability of adequate amount of different nutrients in such material has long been considered as a major constraint. Most of the plant nutrients present in FA do not remain in bio-available forms despite their total occurrence in considerably higher concentrations and take a long time to be transformed to available forms after their application. This behavior has been attributed to inert nature of FA due to restricted microbiological activity in such materials (Chattopadhyay and Bhattacharya, 2000). Under this context, it has been suggested that simultaneous use of FA and organic matter might be an effective proposition for improving this property. It was conceived that use of organic matter would help to extract more nutritional elements to available forms through acid extraction in one hand and encourage microbial degradation of FA to more soluble forms on the other. In fact, many of the significant beneficial effects of FA application in agriculture have been observed when the ash was incorporated with organic materials (Christy and Raghupathy, 2001). Field experiment on use of coal ash in different combinations with organic manure and chemical fertilizers for augmenting crop production and soil productivity in rice based cropping systems has shown increased grain yield in case of wet season rice owing to application of FA in combination with organic materials and chemical fertilizers (Mitra *et al.*, 2000). The physico-chemical properties of the soils also improved in terms of decrease in bulk density, increase in pH, organic carbon and available nutrients viz. N, P, K, Ca, Mg, Zn, Cu and Co. Further, there was improvement in microbial activity under the integrated use of FA and organic materials. Another set of elaborate field trials to assess the effects of different organic and industrial wastes in improving productivity of acid lateritic soil under rice-groundnut cropping system also showed integrated use of FA and organic wastes to be more effective for obtaining grain and pod yields of rice and groundnut (Karmakar *et al.*, 2000).

All these studies indicate that in spite of containing many of the nutrients in high amount, FA cannot provide sufficient amount of plant nutrition through alone application of this material. Most of these nutrient elements in FA remain largely in insoluble forms which are not readily available to plants and may become accessible to them only after due transformation (Chattopadhyay and Bhattacharya, 2010). Hence, for increasing acceptability of FA as a source of plant nutrients, the availability of the nutrients in it needs to be improved substantially. Encouragement of microbial activity for promoting biological as well as chemical degradation of FA and consequent release of various nutrients in to soluble form may be considered as an effective proposition in this regard. Important role of different microorganisms in the development of soils from inert parent materials through various biochemical reactions resulting from decomposition of organic matter especially formation of different acids that dissolve and extract various nutrients from insoluble forms have been discussed by different workers (Brady, 1980; Miller and Donahue, 1992). Similar activities may appear to be useful for improving available nutrient status of FA too. However, a major problem associated with FA is poor occurrence as well as activity of different microorganisms to encourage release of different nutrients from this material, as have been reported by Pitchel and Hayes (1990) and others. Reduction in nitrification potential in FA amended soils under laboratory condition has also been reported (Cervelli *et al.*, 1986). Such behaviors of soil microorganisms may largely be attributed to poor occurrence of organic matter in FA because of the fact that organic matter controls the growth and behavior of most of the microorganisms in any substrate. On the other hand, Schutter and Fuhrman (1999) have reported some enzymatic activities and microbial N cycling functions to increase in FA added soils during a period of 20 months. It has been opined that gradual leaching of soluble salts and trace elements over time helped to ameliorate any initial toxicities of FA allowing plants and microorganisms to respond positively to improved soil conditions. However, the workers also stated that because of increased crop yields in FA amended soils, plant biomass in the soils increased and suggested that this increment in soil organic matter might have eventually enhanced microbial activities in FA treated soils. Such beneficial effects of organic matter on microbial activity and population of FA amended soils have been reported by workers like McCarty *et al.*,(1994) too. They have categorically shown that the populations of different microorganisms like bacteria, actinomycetes and fungi increased considerably due to co-application of FA and sewage sludge in soils as against use of FA alone. Similar benefits have been recorded for the enzymatic activities also.

All these studies lead to the fact that for increasing the nutrient supplying capability of FA, the microbiological activity in this material need to be improved substantially. Since organic matter provides food and energy to different heterotrophic microorganisms and can extract various nutrients into soluble forms from different insoluble minerals through release of some organic acids (Rao and Chhonkar, 1998), simultaneous use of FA and organic materials is likely to emerge as an effective proposition for increasing the efficiency of FA. Countries like India, which have tropical climatic conditions too have vast areas of such soils with low to very low organic carbon content. Simultaneous use of organic materials and FA is likely to result in some beneficial effects on such soil health, which is difficult to achieve through FA alone. However, in view of the escalating quantum of FA generation and also restricted availability of commonly used organic manures, exploration of the possibilities of using the huge resources of different kinds of organic wastes in the country for this purpose appears to be imperative (Chattopadhyay and Bhattacharya, 2010). This approach will not only improve the possibility of recycling FA as a source of plant nutrients but will also help in simultaneous abatement of environmental pollution created by these organic waste materials.

## Possibilities of Recycling Fly Ash through Vermicomposting

In view of the positive effects of application of FA along with organic materials, the potentials for recycling huge resources of both FA and different organic wastes have come into limelight. Integrated use of these organic wastes and FA will not only help to extract more amount of plant nutrients into available forms from insoluble fractions in FA as discussed earlier but also will help to obtain the additional benefits of organic matter on soil health. However, the major problem of recycling these waste materials in agriculture is that, these organic materials can not be applied in agricultural fields unless they are composted properly. Since most of these wastes contain various degradation resistant components, composting of these organic materials often becomes very difficult. It is therefore, adoption of an efficient composting method appeared to be necessary for simultaneous recycling of FA and organic waste.

Vermicomposting biotechnology is recently emerging simple biotechnology and provides better quality compost as compared to traditional methods of composting (Kale, 1993). Epigeic earthworms, which live on the surface of the soils and feed on organic debris, harbor rich concentrations of different kinds of microorganisms, enzymes etc. in their intestines (Edwards and Lofty, 1972; Senapati, 1993). The organic materials excreted by these earthworms in half digested forms, contain these microorganisms and enzymes in high concentrations, which help in rapid aerobic decomposition of these excreted semi digested organic materials resulting in the production of good quality compost in lesser period of time along with multiplication of different beneficial microorganisms in these compost materials (Chattopadhyay, 2004). In view of this efficiency of vermicomposting biotechnology in decomposing different kinds of organic wastes, extensive studies have been carried out in the Soil Testing Laboratory of the Institute of Agriculture, Visva-Bharati University, India, to assess the efficiency of this technology in facilitating plant nutrition through FA. The studies showed considerable increased population of different microorganism's viz. bacteria, actinomycetes and fungi in FA (Figure-2). It has been believed that this behavior helped to encourage the degradation of FA during vermicomposting and promote release of nutrients to available forms (Bhattacharya and Chattopadhyay, 2002).

The findings also support the observations of some yard studies carried out in this laboratory to assess the effects of incubating earthworm, *Eisenia fetida* in different mixtures of FA and cow dung. It was found that this technology helped to maintain higher amount of plant nutrients in available form in the mixtures (Figure-3). It was further observed that for these increments in nutrient availability, FA contributed a good share (Bhattacharya and Chattopadhyay, 2002,04).

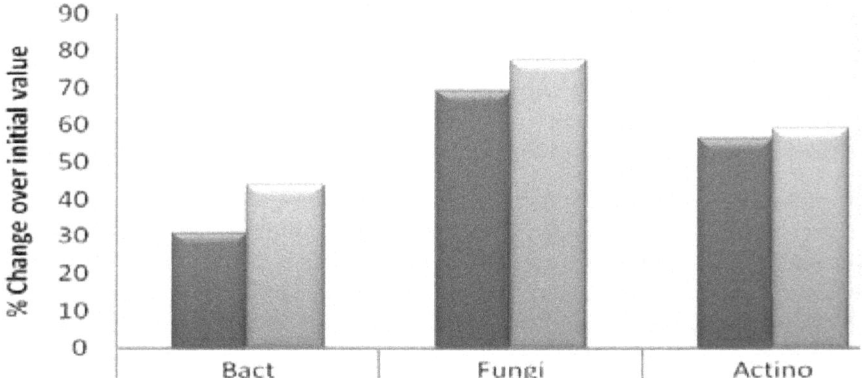


| | Bact | Fungi | Actino |
|---|---|---|---|
| Compost | 30.66 | 68.97 | 56.12 |
| Vermicompost | 43.73 | 76.92 | 58.4 |

**Fig. 2:** Changes in occurrence of different microorganisms in FA: organic waste system due to traditional composting and vermicomposting

The results indicated that during the process of vermicomposting, the microbial organisms involved in the composting process not only decomposed the organic components of the mixture, but also could degrade the FA to a considerable extent and bring a part of insoluble nutrients of fly ash in to more forms that are available. Not only this, the concentrations of nitrogen fixing and phosphorous solubilising bacteria were increased considerably in different organic-ash mixtures on vermicomposting (Bhattacharya and Chattopadhyay, 2002, 2004) (Table-2).

**Table 2: Average occurrence (nos.×$10^8$ $g^{-1}$) of nitrogen fixing bacteria (NFB) and phosphate solubilising bacteria (PSB) in different treatment combinations after 50 days of incubation with and without earthworms**

| Treatment | NFB | | PSB | |
|---|---|---|---|---|
| | E | $E_0$ | E | $E_0$ |
| FA+CD (1:1) | 22.00 | 0.02 | 78.80 | 0.12 |
| FA+CD (1:3) | 1.05 | 0.85 | 4.60 | 0.014 |
| FA+CD (3:1) | 3.15 | 2.63 | 7.50 | 0.068 |

*E = with earthworms, $E_0$ = without earthworms*

*FA+CD= Different ratios of FA and cow dung*

*Source : Bhattacharya and Chattopadhyay (2002, 2004).*

In many cases, occurrences of some heavy metals are considered to be a matter of concern for using FA in agriculture (Sims et al, 1995; Nass et al., 1993). It was interesting to observe that the concentrations of some heavy metals viz. Cr, Cd and Pb declined considerably in the vermicomposted organic-FA mixtures (Bhhattacharya and Chattopadhyay, 2006). This behaviour was attributed to the formation of metallothionein isoforms in earthworms due to exposure to these metals(Bhhattacharya and Chattopadhyay,2006 ; Furst and Nguyen, 1989; Roesijadi,, 1993). Since the earthworms are isolated from the vermicomposted FA prior to the application in fields, use of this technology is likely to reduce the problem of heavy metal toxicity in FA to a considerable extent. Similar effects of using vermicomposting biotechnology on recycling of FA have been reported by other workers too (Saxena *et al.*, 2003; Venkatesh and Eevera, 2008).

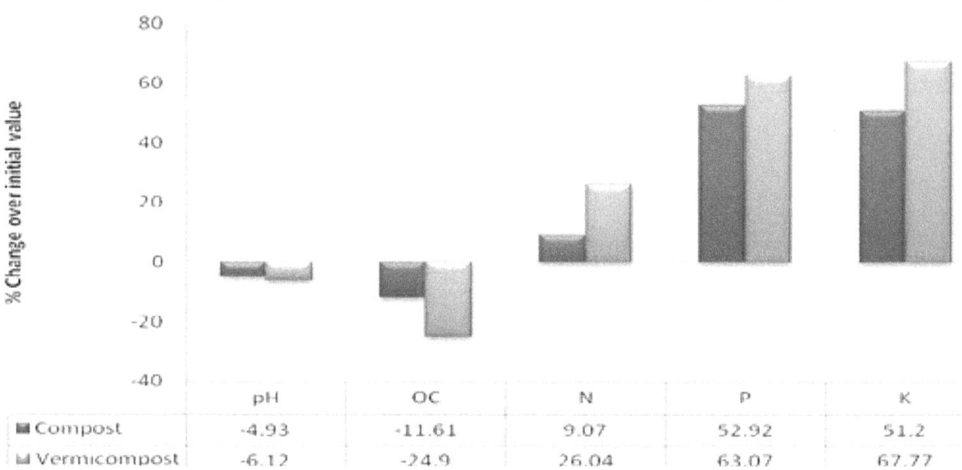


**Fig. 3:** Changes in pH and availability of different plant nutrients in FA: organic waste systems after composting and vermicomposting

*(Source: Roy et al, 2014)*

These beneficial effects of vermicomposting of fly ash have been observed in field level too. Use of vermicomposted fly ash resulted in substantial yield increments in case of potato (Chattopadhyay and Bhattacharya, 2012).. It has been shown that integration of vermicomposted FA with mineral fertilizers played a very important role in producing potato in red and lateritic soils and helped to reduce the requirement of chemical fertilization to nearly 40% level (Table-3).

**Table 3: Potato yield (t ha$^{-1}$) under different treatments using vermicomposted FA in a red- lateritic soil of West Bengal, India (Bhattacharya *et al*, 2012)**

| Treatments | Yield (t ha$^{-1}$) (2009) | Yield (t ha$^{-1}$) (2010) |
|---|---|---|
| Control ($NPK_{100}$) | 31.24c | 31.95d |
| FYM+ $NPK_{100}$ | 35.85a | 36.08b |
| VF+$NPK_{100}$ | 36.29a | 37.20a |
| VF+$NPK_{75}$ | 34.48b | 35.36b |
| VF+$NPK_{50}$ | 31.68c | 34.54c |

*(Values followed by different letters are statistically different)*

*$NPK_{100}$ = 100% recommended NPK fertilizer*

*$NPK_{75}$ = 80% recommended NPK fertilizer*

*$NPK_{50}$ =60% recommended NPK fertilizer*

*FYM = Farm yard manure; V=Vermicomposted FA (both added @10 t ha$^{-1}$)*

Similar observations have also been made in other studies on the effect of vermicomposted FA on integrated nutrient management of potato (Iftikar *et al.*, MS). It has been found that use of vermicomposted FA compared well with farmyard manure and produced comparable yields even after considerable reduction in the dose of chemical fertilizers (Figure - 3).

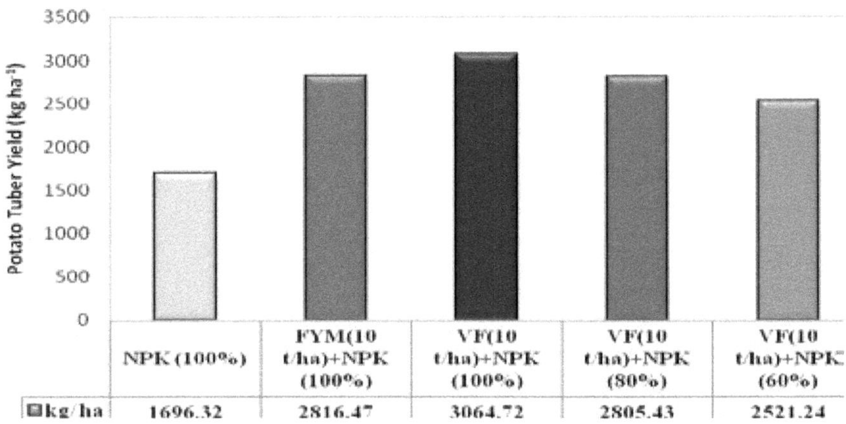


**Fig. 4**: Effect of vermicomposted FA on the yield of potato under integrated nutrient management

*(NPK100,80 and 60% = % of recommended NPK fertilization, FYM= farm yard manure,VF= vermicomposted fly ash)*

Several other farm trials on effect of vermicomposted fly ash on the growth and yield level of crops like tomato, mustard and rice have also shown similar results. A picture of growth of tomato under different treatments has been presented in Figure-4.

Fig. 5: Effect of vermicomposted fly ash on th growth of tomato crop

## Protocol for Preparation of Vermicomposted FA

Based on the knowledge accrued so far, a protocol for production of vermicomposted FA has been prepared at the Institute of Agriculture, Visva-Bharati University. The protocol may be summarized as follows:

i) **Mixing of organic waste with fly ash:** Use of FA and easily degradable organic wastes in 1:1 ratio has shown best results.

ii) **Primary curing of the substrates:** After mixing, the mixtures need to be kept under moist condition for 8-10 days or until the period, thermophillic stage persists.

iii) **Incubation of earthworms:** Generally, epigeic earthworms are to be used for this purpose. Amongst various species of such earthworm, *Eisenia fetida* has been observed to exhibit highest efficiency in terms of survival as well as capability of degrading the wastes.

**iv) Maintaining optimum environmental conditions:** Vermicomposting is an aerobic microbiological process. Hence, maintenance of proper levels of moisture and aeration in the composting system appears to be very important. Studies carried out by the authors have shown a moisture range of 40-50% and maintaining an aerobic condition through frequent turning of the substrates to be conducive for optimization of microbial degradation. The progress of such composting is the maximum under 25-35°C range.

**v) Additional microbial management:** Since FA contains very little amount of nitrogen, inoculation of nitrogen fixing bacteria before the release of earthworms has shown promising results. Similarly, use of phosphate solubilizing bacteria also helps to increase the availability of phosphorus in this composted material.

**vi) Harvest of the product:** Under optimum condition, the vermicomposted fly ash should be ready within 50-60 days of incubation with earthworms. At this stage, the materials are air dried and sieved. The composted materials pass through the sieve and the worms are trapped above the sieve. These are then collected and re-used for the next set of vermicomposting.

## Conclusion

Vermicomposting may be considered as an effective tool for facilitating the re-use of inorganic waste like fly ash in agriculture. Good amount of insoluble nutrients of FA may be transformed into bioavailable form through this bioprocessing. However, this is a new field of study and more information need to be collected on different aspects of this biotechnology

## References

Adriano, D.C., Page, A. L., Elseewi, A. A., Chang, A. C. and Straughan, I. ( 1980): Utilization and Disposal of fly ash and other coal residue in terrestrial ecosystems – a review. *J. Environ. Qual.*9: 333-334

Bhattacharya, S.S. and Chattopadhyay, G.N.(2001): Vermicomposting as a tool for increasing availability of nutrients in fly ash. In: Agricultural Application of Fly Ash. Proc. II Nat. Sem. On Use of Fly Ash in Agriculture. Annamalai University, Annamalainagar, India. p.128-132.

Bhattacharya, S.S. and Chattopadhyay, G.N., (2002):Increasing bioavailability of phosphorous from fly ash through vermicomposting. *J. Environ. Qual.* 31: 2116-2119.

Bhattacharya, S.S. and Chattopadhyay, G.N., (2004): Transformation of nitrogen during vermicomposting in fly ash. *Waste Management & Research* 22(6):488-491.

Bhattacharya, S.S. and Chattopadhyay, G.N., (2006): Effect of vermicomposting on the transformation of some trace elements in fly ash. *Nutr. Cycl. Agroecosyst.* 75:223-231.

Bhattacharya, S.S., Iftikar, W., Sahariah, B. and Chattopadhyay, G.N. (2012): Vermicomposting converts fly ash to enrich soil fertility and sustain crop growth in red and lateritic soils. *Resources, Conservation and Recycling* 65:100-106

Brady, Nyle, C. (1980): Nature and Properties of Soils. Eurasia Publ. House (P) Ltd. New Delhi. p. 339

Cervelli, S.G., Petruzzelli, A. Perna and Menicaglis, R.(1986):Soil nitrogen and fly ash utilization-a laboratory investigation. *Agrochemica* 30: 27-33.

Chattopadhyay, G.N. (2004): Vermicomposting as a biotechnological tool for recycling organic wastes. In: Biotechnology in Environmental Management T.K. Ghosh, T. Chakrabarti and G. Tripathi (Ed.).Vol. I. A.P.H. Publishing Corporation, New Delhi, p.135-145.

Chattopadhyay, G.N. and Bhattacharya, S.S.(2000): Possibility of increasing available nutrient status of fly ash through vermicomposting. Proc. II Int. Conf. on Fly Ash Disposal and Utilization, New Delhi.Vol-2. p. 14-16.

Chattopadhyay, G.N. and Bhattacharya, S.S. (2010: Use of Coal Ash in Agriculture. Bull. Coal Ash Institute of India. Kolkata, India.

Christy, N.M.P. and Raghupathy, B. (2001): Effect of lignite fly ash in an integrated nutrient management of rice cv ADT 38 on magnesium content and uptake in backfield areas of Neyveli open cast mines. In: Agricultural application of fly ash.Proc.2nd Nat. Sem. on Use of Fly Ash in Agriculture. Annamalai University, Annamalainagar.India. p. 125-127

Edwards, C.A. and Lofty, J.R. (1972): Biology of Earthworms. Chapman and Hall, London.

Furst, A. and Nguyen, Q. (1989): Cadmium induced metallothionein in earthworms (*Lumbricus terrestris*). *Biol. Trace Elem. Res.* 21: 81-85.

Garau, M. A., Dalmau, J.L. and Felipo, M.T. (1991): Nitrogen mineralization in soil amended with sewage sludge and fly ash. *Biol. Fertil. Soils.*12: 199-201.

Hill, M.F. and Lamp, C.A. (1980): Use of pulverized fuel ash from Victorian brown coal ash as a source of nutrients for a pasture species. *Australian J. Exp. Agric. Anim. Husb.*, 20: 337-384.

Iftikar, Wasim; Roy, Gourab and Chattopadhyay, G.N. (MS): Integration of vermicomposted fly ash in nutrient management of potato in a red and lateritic soil. Personal communication.

Kale, Radha, D. (1993): Earthworms: The biological tools for healthy and productive soils. Congress on Traditional Sci. Tech. India. IIT, Bombay.

Karmakar, S., Mitra, B.N. and Ghosh, B.C. (1997): 1st All India Peoples Technology Congress, Calcutta, India.

Kumar, V.; Goswami, G. and Zacharia, K.A. (1999): Fly ash: its influence on soil properties. Workshop, Indian Society of Soil Science, 18-21st October, Calcutta.

Kumar, V.; Zacharia, K.A. and Goswami, G. (2000): Fly ash use in Agriculture: a perspective. In: Proc of 2nd Int. Conf. on fly ash disposal and utilization (Fly Ash II), 2-4 Feb, New Delhi, India, pp-IX-1-13.

Kumari, K. and Dubey, P.S. (2003): Effect of fly ash on growth and nutritional status of spinach.In: Proc. of 3rd Int. Conf. on Fly Ash Disposal and Utilization (Fly Ash II),19-21 Feb., New Delhi, India, p. VII-54-60.

McCarty, G. W. R.; Siddarmappa, R. J. ; Wright, E. E. and Gao, G. (1994): Evaluation of coal combustion byproducts as soil liming materials: Their influence on soil pH and enzyme activities. *Biol. Fertil. Soils*.17: 147-172.

Miller Raymond, W. and Donahue Roy, L.(1992): Soils - An Introduction to Soils and Plant Growth. Prentice Hall of India Pvt. Ltd., New Delhi. p.768.

Mitra, B. N.; Ghosh, B. C.; Karmakar, S.; George, J.; Rautray, S. K. and Basu, S. (2000): Utilization of industrial solid wastes in augmenting crop production and soil productivity .In: B. B. Jana, R. D. Banerjee, B. Guterstam and J. Heeb (Eds.). Waste Recycling & Resource Management in the Developing World. p. 185-195.

Mulford, F. R. and Martens, D.C. (1971): Response of alfalfa to boron in fly ash. *Soil Sc. Soc. Am. Proc.* 35: 296-300.

Nass, M. M.; Lexmond, Th. M.; Beusichem, van, M. L. and Jurkovicova, M. Janessem (1993): .Long term supply and uptake by plants of elements from coal fly ash. *Commun. Soil Sci. Plant Anal.* 24 (9 & 10): 899-913.

Page, A.L.; Elseewi, A. A., Lund, L. J. Bradford, G. R., Mattigod, S., Chang, A.C. and Bingham, F.T. (1980): Consequences of Trace Element Enrichment of Soils and Vegetation from the combustion of Fuels Usedin Power Generation. University of California, Riverside, 158.

Pitchel, J.R. and Hayes, J. M. (1990): Influence of fly ash on soil microbial activity and population *J. Environ. Qual.* 19: 593-597.

Rao, D.L.N. and Chhonkar, P.K. (1998): Organic matter in relation to soil biological quality and sustainability In: T.D. Biswas and G. Narayanaswamy (Eds.) Soil Organic Matter and Organic Residue Management for Sustainable Productivity. Indian Soc. Soil Sci. New Delhi. p. 80-89.

Roesijadi, G. (1993): Response of invertebrate metallothioneins and MT genes to metals and implications for environmental toxicology. In: K.T. Suzuki; N. Imura and M. Kimura (Eds). Metallothionein III. Birkhauser Verlag, Switzerland. p. 141-158.

Roy, Gourab; Iftikar Wasim and Chattopadhyay, G.N. (2014): Use of vermicomposting biotechnology for recycling fly ash as a source of plant nutrients. In: Chaudhuri and Singh (Eds): Biology and Ecology of Tropical Earthworms. Discovery Publ. House Pvt. Ltd. New Delhi. p. 230-239.

Roy. Gourab; Saha, Abhisek; Iftikar, Wasim; Chattopadhyay, G. N. and Mandal, B. (2009): Effect of earthworm on recycling of fly ash for agricultural use. Int. Conf. Recent trends in life science researches vis-à-vis natural resource management, sustainable development and human welfare. June 27-29. Hazaribagh, India.

Saxena, M. and Asokan, P. (2000): Effect of coal ash (NTPC, Rihand Nagar) on soil fertility and various agricultural products. In: Proc. of 2nd Int. Conf. on Flyash Disposal and Utilization (Fly Ash II), 2- 4 Feb., New Delhi, India, p. IX-47-62.

Saxena, M.; Murali, S.; Asokan, P.; Mishra, C.R. and Pal, H.K (2003): Safe and bulk utilization of Nalco pond ash in agriculture. In: Proc. of 3rd Int. Conf. on flyash disposal and utilization (Fly Ash II),19-21 Feb, 2003, New Delhi, India, p-VII-45-53.

Schutter, Macy, E. and Fuhrman, Jeffry (1999): Microbial responses to coal fly ash under field condition. *J. Environ Qual.* 28: 648-652.

Senapati, B.K. (1993): Earthworms gut content and its significance. In: A. K. Ghosh (Ed.). Earthworm Resource and Vermiculture, Z.S.I., Calcutta, India, p. 97-100.

Sims, J. T.; Vasilas, B. L. and Ghodrati, M. (1995): Evaluation of flyash as soil amendment for the Atlantic coastal plain : II Soil chemical properties and crop growth. *J. Water Soil Pollut.*, 81:363-372.

Suzuki, K.T., Yamamura, M., Mori, T., (1980) : Cadmium-binding proteins induced in the earthworm. *Arch. Environ. Contam. Toxicol.* 9(4): 415-424.

Tripathi, Prem S.M.; Singh, Gulab and Tripathi, Ramesh, C. (1997): 1st All India Peoples Technology Congress, Calcutta, India

Venkatesh, R.M. and Eevera, T. (2008): Mass reduction and recovery of nutrients through vermicomposting of fly ash. *Appl. Ecol. Environ. Res.* 6 (1):77-84.

Wong, J. W. C. and Wong, M. H. (1987) : Co-recycling of fly ash and poultry manure in nutrient deficient sandy soil. *Resource and Conserv.*13: 291-304.

# 7

# Organic Agriculture - Option to Save Soil Health from Persistent Pollutants

*Deepmala Satpathy and M.Vikram Reddy**

*Department of Ecology and Environmental Sciences, Pondicherry University, Pudicherry- 605 014 (India)*

## Soil and Persistent Pollutants

Soil is a dynamic natural body on the surface of earth. It has four major components, *i.e.* mineral matter, organic matter, soil water and soil air. The amount of these components in the soil varies with the locality. Presence of different kinds of heavy metals such as Cd, Cu, Mn, Bi and Zn etc. in trace or in minimum level is a natural phenomenon but their increased level is an indicator of the degree of pollution load in that specific area. Modern agricultural practices have polluted the soil to a large extent. Today with the advancing agro-technology, huge quantities of fertilizers, pesticides, herbicides, weedicides and soil conditioning agents are employed to boost up the crop yield. Many agricultural lands have now excessive amounts of plants and animal wastes which are posing the problem of soil pollution.

Soil pollution with heavy metals has become a vital environmental concern due to their potential adverse effects. Heavy metals are considered as soil contaminants owing to their wide-ranging present. These are non-biodegradable, non thermo-degradable and hence readily build up to their toxic levels which causes adverse effects in whole ecosystem. Heavy metal pollution can remain dormant for a long time and then surface with vengeance. These might affect the total biosphere for a longer time since these are non-degradable. It is known that heavy metals form an important polluting group. These have not only toxic and carcinogenic effects but also tend to accumulate in living organisms and cause various diseases (Chopra *et al.*, 2009).

Soil is considered to be an excellent medium to monitor and assess heavy metal pollution because anthropogenic heavy metals are usually deposited in the surface soils (Govil *et al.*, 2002). Heavy metal contaminated soil adversely affects the whole ecosystem when these toxic heavy metals migrate into groundwater as runoff or taken up by flora and fauna, which may result in great threat to whole ecosystem due to translocation and bioaccumulation (Bhagure and Mirgane, 2011). Heavy metals become potentially toxic to crop plants, animals, and human beings when the contaminated soils are used for agricultural production (Wong *et al.*, 2002). These have a hazardous effect not only on crop plants but also on human health. Environmental contamination of the biosphere with heavy metals due to intensive agricultural and other anthropogenic activities poses serious problems for safe use of agricultural land (Fytianos *et al.*, 2001).

The specific knowledge of these kinds of heavy metals, their sources, forms and dependence on soil provides an authentic idea for better soil management practices. Remediation of heavy metal contaminated soils is necessary to reduce the associated risks, make the land fertile and resource available for agricultural practices which can ultimately enhance food quality and security. It has recently been noticed that even the use of organic fertilizers is not much safe to improve soil health as these also add persistent pollutants into the soil system. The present article is the outcome of the findings in the level of increment of heavy metals after the use of inorganic fertilizers in the agrarian soil. The article will also be useful to the farmers in choosing specific organic or inorganic fertilizers that might add the minimum level of such persistent pollutants in the soil than the others for sustainable farming.

## Sources and Risks of Soil Heavy Metal Contamination

The environmental impacts of heavy metals always go unnoticed and are rarely catastrophic to merit immediate measures. Except for recent report on arsenic poisoning in West Bengal, no major episode of metal toxicity has been recognized in India. In addition to air and water, about 6,209,000 tons of heavy metals enter the soil every year through various important sources like: (a) Agricultural and animal waste (b) Municipal and industrial sewage (c) Fertilizers and pesticides (d) Atmospheric fallout. Soil contamination should be a primary concern in India because the country relies heavily on agriculture. Agricultural lands in many industrial areas have already been rendered useless because of the presence of high concentrations of heavy metals like Pb, Zn, Cr, Cu, Fe etc. Huge scale application of fertilizers and pesticides has also contaminated the soil with large amounts of heavy metals, such as Cd, Pb and Zn (Sharma 2006).

Major sources of soil heavy metals contamination are irrigation by discharged industrial effluents, sewage and waste water irrigation, waste disposal etc. Gupta *et al.* (2007) found that leather industries located at Jajmau near Kanpur city are the major sources of heavy metal contaminations in the agricultural soil in the surrounding areas where treated effluents have been used for irrigation. Rattan *et al.* (2005) have reported that Keshopur effluent irrigation scheme in Delhi resulted in significant build up of DTPA extractable Zn (208%), Cu (170%), Fe (170%), Ni (63%), and Pb(29%) in sewage irrigated soils. Surface soil gets accumulated with toxic heavy metals such as Cd, Zn, Cr, Ni, Pb and Mn when irrigated with waste water. The long-term use of sludge can cause heavy metal accumulation in soils (López-Mosquera *et al.*, 2000). Heavy metals can also contaminate the food chain and reduce crop yields (Obrador *et al.*, 1997 and Wang *et al.*, 2003). The wastewater irrigation practice is widely adopted by resource poor farmers but the potential health risks are often ignored. Waste water is mainly contaminated with industrial effluents that contain high levels of heavy metals. Hence dietary intake of heavy metals through the harvested crops is a substantial risk to the health of families who depend upon the use of contaminated irrigation water for growing crops to meet their food requirements (Chopra *et al.*, 2009).

## Soil Contamination Due to Use of Agrochemicals

Heavy metals enter into agro ecosystem through both natural and anthropogenic processes. Soil inherits trace elements from its parent materials. Heavy metals in soil are derived from both parent materials and anthropogenic inputs. Anthropogenic processes include use of inorganic fertilizers, pesticides, sewage and waste water irrigation etc. These processes result in enrichment of heavy metals in agro ecosystem. The mobility and availability of trace elements are controlled by many chemical and biochemical processes such as precipitation–dissolution, adsorption–desorption, complexation-dissociation and oxidation–reduction. It is therefore, essential to understand some major reactions in soils that control the release of a specific heavy metal in the soil and the environment in order to overcome problems related to deficiency and contamination of these elements. (He *et al.*, 2005). Accumulation of heavy metals in the soil has potential to restrict the soil's function, cause toxicity to plants, and contaminate the whole food chain.

### Fertilizers, Pesticides, Manures and Chemicals

Contamination of agricultural soil is mainly related to input and accumulation of heavy metals due to repeated use of inorganic fertilizer, farm manures, fungicides, pesticides and bio-solids (Zhang *et al.*, 2010). Chemical fertilizers, fungicides and pesticides containing Cu, Zn, and As have been widely used to protect various fruits and vegetables which are the main reason for heavy metal accumulation in soil. Large quantities of fertilizers are regularly added to soils in intensive farming systems to provide adequate N, P, K for crop growth. The compounds used to supply these elements contain trace amounts of heavy metals (e.g. Cd and Pb) as impurities which after continued application of fertilizers may significantly increase their content in the soil (Jones and Jarvis, 1981). Metals such as Cd and Pb have no known physiological activity. Repeated use of P fertilizers such as triple superphosphate may result in accumulation of these elements and increases the contamination potential especially of Cd in the soil. (He *et al.*, 2005).

Chemical fertilizers contain trace amounts of heavy metals. Varying concentrations of Cd has been found in Phosphorus (P) fertilizers such as triple superphosphates and calcium/magnesium phosphate (He *et al.*, 2005). Some P fertilizers can have Cd concentrations up to 450 mg kgl and hence they are banned for use in agriculture in many countries (Mortvedt *et al.*, 1995). In USA the trace elements such as Cu, Zn, Fe, and Mn were purposely added into regular blend fertilizers to meet the demand of plant growth for these elements. These fertilizers are important sources of trace elements for crops growing in soils subject to severe deficiency in these elements such as sandy, peaty and calcareous soils (He *et al.*, 2005). Organic materials such as farm manures, bio-solids or composts are also found to contain higher concentrations of trace elements. Use of bio-solids and composts has been reported to increase total amounts of Cu, Zn, Pb, Cd, Fe, and Mn in the soils (McBride, 2004).

## Irrigation

The input of metals through irrigation varies noticeably from place to place. Non-contaminated fresh and salt water contains extremely low concentrations of heavy metals than domestic and industrial wastewaters. Repeated use of wastewaters that have not been treated to filter heavy metals may significantly contribute to the accumulation of these toxic heavy metals in the soil.

## Outputs of Heavy Metals from Agro Ecosystems

Outputs of heavy metals from agro ecosystems include crop harvest, losses by surface leaching, surface runoff and gaseous emission. Crop harvest accounts for a big proportion of the output of these toxic heavy metals although the exact amounts of metal removal vary significantly with soil texture, crop variety and climate conditions (He *et al.*, 2005). Surface runoff losses of heavy metals are associated with transport of particulate matters that contain adsorbed heavy metals. Gaseous losses are meant for heavy metals such as As, Se and Hg. Some volatile organic like Se and As compounds are formed in plants and also get emitted from plant leaves. Hg can be converted into methyl-Hg and emitted from the soil through microbial activity (He *et al.*, 2005).

## Contamination of Agricultural Soils Due to Heavy Metals- A Case Study

Several studies are being conducted to examine the contamination of agricultural soils due to heavy metals. However little information is available about heavy metal contamination of agricultural land in India. A study was done for assessment of potential toxic and non-essential heavy metals in the surface soil of paddy fields at the predominantly paddy-cultivated area nearby Kalpakkam in Tamil Nadu state (India) (Satpathy *et al.*, 2014).

**Table 1: Mean values of heavy metals (Pb, Cd, Cu, Cr, Mn and Zn) for uncontaminated paddy soils, mean values for normal surface soils worldwide, critical concentrations for contaminated soils, Indian and European standards compared with the values of the present study.**

| Elements (µg g-1) | Mean values for paddy soils[a] | Mean values for normal surface soils world wide[b] | Critical soil concentration[c] | Indian standard [d] | European Union standards[e] (EU 2002) | Kalpakkam area[f] |
|---|---|---|---|---|---|---|
| Pb | 23.3 | 22-44 | 100-400 | 250-500 | 300 | 5.3-19.8 |
| Cd | 0.34 | 0.37-0.78 | 3-8 | 3-6 | 3 | 0.2-0.6 |
| Cu | 20.7 | 13-24 | 60-125 | 135-270 | 140 | 1.10-2.9 |
| Cr | 64 | 12-83 | 75-100 | - | - | 1.3-7.8 |
| Mn | 0.39 | 0.27-0.53 | 1.5-3 | - | - | 12.5-33.9 |
| Zn | 61 | 45-100 | 70-400 | 300-600 | 300 | 3.8-33.8 |

[a] *Mean of total concentration of elements in uncontaminated paddy soils. Data are from Wong et al. (2002); Wang et al. (2003); Chandrajith et al. (2005).*

[b] *Mean of total concentrations of elements in the surface horizon of normal worldwide soil, normal mean that the data do not include contaminated or mineralized soils. Data are from Kabata–Pendias (2001); Essington (2004).*

[c] *Higher concentrations may be toxic to plants depending on speciation (Alloway, 1995).*

[d] *Indian standards (Awashthi, 2000) for agricultural soils*

[e] *European standards (EU 2002) for agricultural soils*

[f] *Mean concentration of elements in paddy soils of Kalpakkam area. Data are from Satpathy et al. (2014)*

The indiscriminate use of agro-chemicals, such as chemical fertilizers and pesticides probably resulted in elevated concentrations of the heavy metals in the soil in paddy fields. Some of the heavy metals have potency to leach into deeper soil layers, which may affect the ground water. Hence the food chain of the agro-ecosystem may get contaminated due to the intensive agricultural practices. The paddy plants including grains are also found to accumulate the heavy metals in higher concentration causing deleterious health effects of human beings who depend on rice up to 70 % especially in tropical countries including India (Reddy *et al.*, 2013).

## Soil Management

Soil and crop management methods can help in preventing the uptake of the toxic heavy metals by the plants. The soil acts as major sink through which the toxic heavy metals exert their toxic effects (Brady and Weil, 1999). Heavy metals get immobilized with increase in soil pH and peaks under mildly alkaline conditions.

Heavy metals mobility is associated to their immobilization in the solid phase (Fuller, 1977). Certain management practices will not only help in removing heavy metal contaminants, but will also help to immobilize them in the soil and reduce the potential of adverse effects on the ecosystem (Chopra *et al.,* 2009). Cationic metals are known to be more soluble at lower pH level and therefore increasing the pH in soil makes them less available to plants and consequently less likely to be incorporated in their tissues and ingested by humans. Drainage improves soil aeration and allows metals to oxidize, making them less soluble. Therefore when aerated, these metals are less available. At the same time chromium is more available in the oxidized forms so active organic matter helps in reducing the availability of chromium in the soil. Heavy phosphate application in the form of phosphate fertilizers reduces the availability of cationic metals but have the opposite effect on anionic compounds like arsenic which has a great adverse impact on then human health as it is very toxic in nature. It is for this reason the use of phosphate fertilizers should be checked (Chopra *et al.,* 2009). Several techniques for management of contaminated soil such as phyto-remediation, bioremediation and vermi-remediation are being encouraged for detoxification of the contaminated soils which has become very necessary for a healthy environment in the present scenario.

## Plants in Management of Metal Contaminated Soil

Plants have been used to stabilize or remove metals from soil and water. It has been confirmed that plants are effectual in cleaning up contaminated soil (Wenzel *et al.,* 1999). Phyto-remediation is a technique which involves plants to remove or degrade different kinds of soil pollutants such as heavy metals, pesticides, solvents, crude oil, poly-aromatic hydrocarbons etc. These heavy metal accumulating plants are known as hyper-accumulators. In phyto-remediation process, accumulation and distribution of heavy metals in plant tissues are important aspects to evaluate the role of plants in remediation of contaminated soil (Friedland, 1889). Wildflowers have been used to degrade hydrocarbons from an oil spill in Kuwait. Hybrid poplars can effectively remove toxic compounds such as TNT as well as high concentrations of nitrates and pesticides (Brady and Weil, 1999), through food chain (Pichtel *et al.,* 2000, Chopra *et al.,* 2009). Recently, phyto-extraction method is being accepted where plants are used to extract toxic heavy metals from contaminated soils. This process has emerged as a cost effective, environment-friendly cleanup alternative.

## Vermicompost: A Soil Conditioner and Plant Growth Promoter

Inorganic fertilizers and pesticides are known to contaminate soil with toxic heavy metals which brings adverse effects to soil condition. The toxic heavy metals are then subsequently taken up by plants and enter the food chain, thus posing a potential threat to the pristine nature of the ecosystem as well as the human beings (Taylor and Percival 2001, Parkpian *et al.,* 2003, Reddy *et al.,* 2013). Hence nowadays farmers have started applying organic fertilizers like vermicompost instead of chemical fertilizers.

Earthworm's vermicompost is proving to be highly nutritive 'organic fertilizer' and more 'growth promoter' over the conventional composts and a 'protective' farm input (increasing the physical, chemical and biological properties of soil, restoring and improving its natural fertility) against the chemical fertilizers which destroys soil properties and decreases its fertility over the years (Sinha *et al.*, 2009).Vermicompost is rich in NKP (nitrogen 2-3%, potassium 1.85-2.25% and phosphorus 1.55-2.25%), micronutrients, beneficial soil microbes and also contain 'plant growth hormones and enzymes. It is scientifically proved 'miracle growth promoter and plant protector' from pests and other plant diseases. Vermicompost retains nutrients for long time while the conventional compost fails to deliver the required amount of macro and micronutrients to plants in shorter time (Sinha *et al.*, 2009).

There have been several reports that worm worked waste and their excretory products (vermicastings) can bring excellent plant growth (Krishnamoorthy and Vajranabhaiah, 1986; Tomar *et al.*,1998; Atiyeh *et al.*, 2000a, 2000b; Edwards *et al.*, 2004; Arancon *et al.*,2006; Azarmi *et al.*, 2008). It has been found to influence on all yield parameters such as-improved seed germination, enhanced rate of seedling growth, flowering and fruiting of major crops like wheat, paddy, corn, sugarcane, tomato, potato, brinjal, spinach, grape and strawberry as well as of flowering plants like petunias, marigolds, sunflowers etc (Sinha *et al.*, 2009).

## Conclusion

Chemical fertilizers and pesticides are indiscriminately used by farmers which are probably the main sources of the toxic heavy metals accumulated in agricultural fields. Organic agriculture with little use of agro-chemicals could be the alternative solution for reducing the contamination of toxic heavy metals particularly the Cd, Cr, and Pb in the agricultural fields. Besides, discharges of industrial effluents and sewage on soil are the major factors for heavy metal contamination. There is an urgent need to prevent disposal of effluents on agricultural and forest lands as it may cause severe human health hazard if it is translocated in the food products. Knowledge of the sources, chemistry and potential risks of toxic heavy metals in contaminated soils is necessary for the selection of appropriate remedial measures. Remediation of soil contaminated by heavy metals is necessary in order to reduce health risks, better resource availability and enhance food security. Immobilization, phyto-remediation and use of organic compost like vermicompost are best available methods for cleaning up heavy metal contaminated soils. Vermicompost can be used as manure and plant growth promoter which can also protect plants against pest and diseases as it has no adverse impact on agricultural land as well as agricultural products. However, it was found that even application of organic fertilizers also contaminate the agricultural soil with persistent pollutants like heavy metals and of not persistent organochlorine chemicals like insecticidal compounds (Reddy, 1985)

## References


Alloway, B. J. (1995): The origins of heavy metals in soils, (ed. B. J. Alloway), Chapman & Hall, London, UK. p. 29-39.

Arancon, N.Q., Edwards, C.I., & Bierman, P. (2006): Influences of vermicompost on field strawberries-2: Effects on soil microbiological and chemical properties. *Bioresource Technology*, 97: 831-840.

Atiyeh, R.M., Dominguez, J., Sobler, S. & Edwards, C.A. (2000a): Changes in biochemical properties of cow manure during processing by earthworms *(Eisenia andrei)* and the effects on seedling growth, *Pedobiologia*, 44: 709-724.

Atiyeh, R.M., Subler, S., Edwards, C.A., Bachman, G., Metzger, J.D., & Shuster, W. (2000b): Effects of vermicompost and composts on plant growth in horticultural container media and Soil. *Pedobiologia*, 44: 579-590.

Awashthi, S. K. (2000): Prevention of Food Adulteration Act No. 37 of 1954. Central and State rules as amended for 1999, Ashoka Law House, New Delhi, India.

Azarmi, R., Giglou, M.T., & Taleshmikail, R.D. (2008): Influence of vermicompost on soil chemical and physical properties in tomato (*Lycopersicum esculentum*) field, *African J. Biotechnology*, 7 (14): 2397-2401.

Bhagure, G. R. & Mirgane, S. R. (2011): Heavy metal concentrations in ground waters and soils of Thane Region of Maharashtra, India. *Environmental Monitoring and Assessment*, vol. 173, p.- 643–652.

Brady, N.C., & Weil, R.R. (1999): The nature and properties of soils. 12th ed. Prentice Hall. Upper Saddle River, NJ.

Chandrajith, R., Dissanayake, C. B. & Tobschall, H. J. (2005): The abundances of rarer trace elements in paddy (rice) soils of Sri Lanka, *Chemosphere*, vol. 58, no. 10: 1415–1420.

Chopra, A. K., Pathak, C. & Prasad G. (2009): Scenario of heavy metal contamination in agricultural soil and its management, *J. Appl. Nat. Sci.*, 1:99-108.

Edwards, C.A., Domínguez, J. & Arancon, N.Q. (2004): The influence of vermicompost on plant growth and pest incidence. In: Soil Zoology for Sustainable Development in the 21st Century (ed. Shakir, S.H. and W.Z.A. Mikhail), Self-Publisher, Cairo, Egypt.:397-420.

Essington, M. E. (2004): Soil and Water Chemistry: An Integrative Approach, CRC Press, Boca Raton, Fla, USA.

Friedland, A.J. (1989): The movement of metals through soils and ecosystem. In: Shaw, A.J. (ed.). Heavy metal tolerance in plants: Evolutionary aspects, CRC press, Inc, Boca Raton: 7-20.

Fuller, W.H. (1977): Movement of selected metals, asbestos and cyanide in soil: Application to waste disposal problem, EPA- 600/2-77-020. Solid and hazardous waste research division, U.S. Environmental protection agency, Cincinnati, OH.

Fytianos, K., Katsianis, G., Triantafillou, P. & Zachariadis, G. (2001): Accumulation of heavy metals in vegetables grown in an industrial area in relation to soil. *Bull. Environ. Contam. Toxicol.* vol. 67, no. 3: 423–430.

Govil, P. K., Reddy, G. L. N. & Krishna, A.K. (2002): Contamination of soil due to heavy metals in the Patancheru industrial development area, Andhra Pradesh, India, *Environ. Geol.*, vol. 41: 461–469.

Gupta, A.K., Sinha, S., Basant, A. & Singh, K.P. (2007): Multivariate analysis of selected metals in agricultural soil receiving UASB treated tannery effluent at Jajmau, Kanpur (India), *Bull. Environ. Contam. Toxicol.* 79: 577- 582.

He, Z. L., Yang, X. E. & Stoffella, P. J. (2005): Trace elements in agro ecosystems and impacts on the environment. *J. Trace Elements in Med. Biol.* 19: 125–140.

Jones L. H. P. & Jarvis, S. C. (1981): The fate of heavy metals, in the chemistry of soil Processes (ed. D. J. Green and M. H. B. Hayes). John Wiley & Sons, New York, NY, USA:593.

Kabata-Pendias, A.( 2001): Trace Elements in Soils and Plants, CRC Press, Boca Raton, Fla, USA,

Krishnamoorthy, R.V. & Vajranabhaiah, S.N. (1986) : Biological Activity of Earthworm Casts: An Assessment of Plant Growth Promoter Levels in the Casts, Proc. Indian Acad. Sci. (Animal Science), 95: 341-351.

Lopez-Mosquera, M.E., Moiron, C. & Carral, E. (2000): Use of dairy industry sludge as fertilizer for grasslands in northwest Spain: Heavy metal level in the soil and plant. *Resource, Conservation and Recycling,* 30: 95- 109.

McBride, M.B. (2004): Molybdenum, sulfur, and other trace elements in farm soils and forages after sewage sludge application, *Communications in Soil Science and Plant Analysis,* 35:517–35.

Mortvedt, J.J., & Beaton, J.D. (1995): Heavy metal and radionuclide contaminants in phosphate fertilizers. In: Tiessen H, editor. Phosphorus in the global environment: transfer, cycles and management, New York: Wiley. p. 93–106.

Obrador, A., Rico, M.I., Mingot, J. I. & Alvarej, J.M. (1997): Metal mobility and potential bioavailability in organic matter-rich soil-sludge mixtures: Effect of soil type and contact time, *The Science of the Total Environment,* 206: 117-126.

Parkpian, P., Leong, S. T., Laortanakul, P. & Thunthaisong, N. (2003): Regional monitoring of lead and cadmium contamination in a tropical grazing land site, Thailand. *Environmental Monitoring and Assessment,* 85(2): 157– 173.

Pichtel, J.K., Kuroiwa and Sawyerr, H.T.(2000): Distribution of Pb, Cd and Ba in soils and plants of two contaminated sites. *Environmental Pollution,* 113: 35-40.

Rattan, R.K., Datta, S.P., Chhonkar, P.K., Suribabu, K., and Singh, A.K. (2005): Long term impact of irrigation with sewage effluents on heavy metal content in soils, crops and groundwater- a case study. *Agriculture, Ecosystem and Environment,* 109: 310-322.

Reddy, M. V., Satpathy, D. and Dhiviya, K. S. (2013): Assessment of heavy metals (Cd and Pb) and micro-nutrients (Cu, Mn, and Zn) of paddy (*Oryza sativa* L.) field surface soil and water in a predominantly paddy-cultivated area at Pudicherry (Pondicherry, India), and effects of the agricultural runoff on the elemental concentrations of a receiving rivulet. *Environmental Monitoring and Assessment*, vol. 185: 6693–6704.

Satpathy, D., Reddy, M. V. and Dhal, S.P. (2014): Risk Assessment of Heavy Metals Contamination in Paddy Soil, Plants, and Grains (*Oryza sativa* L.) at the East Coast of India, *BioMed Research International*. http://dx.doi.org/10.1155/2014/545473.

Sharma, B.K. (2006): Environmental chemistry, Goel publishing house, Meerut.

Taylor, M. D. (1997): Accumulation of cadmium derived from fertilizers in New Zealand soils, *Science of the Total Environment*, 208(1- 2): 123–126.

Sinha, R.K., Herat, S., Valani, D. and Chauhan, K. (2009: Earthworms Vermicompost: A Powerful Crop Nutrient over the Conventional Compost and Protective Soil Conditioner against the Destructive Chemical Fertilizers for Food Safety and Security. *American-Eurasian J. Agricul. Environ. Sci.,* 5 (S): 01-55.

Tomar, V.K., Bhatnagar, R.K. and Palta, R.K. (1998): Effect of Vermicompost on Production of Brinjal and Carrot. Bhartiya Krishi Anushandhan Patrika (Indian Agricultural Research Bulletin), 13 (3-4): 153-156.

Wang, Q.R., Cui, Y.S., Liu, X.M., Dong, Y.T. and Christie, P. (2003): Soil contamination and plant uptake of heavy metals at polluted sites in China, *J. Environ. Sci Healt,*.38: 823- 838.

Wang, X. C., Yan, W. D., An et al., Z. (2003): Status of trace elements in paddy soil and sediment in Taihu Lake region. *Chemosphere*, vol. 50(6): 707–710.

Wenzel, W.W., Adriano, D.C., Salt, D. and Smith, R. (1999): Phytoremediation: A plant-microbe based remediation system. In: Bioremediation of contaminated soils (ed. D.C. Adriano *et al.*). American Society of Agronomy, Madison, WI. p. 457-508.

Wong, S. C., Li, X. D., Zhang, G., Qi, S. H. and Min, Y. S. (2002: Heavy metals in agricultural soils of the Pearl River Delta, South China, *Environmental Pollution*, vol. 119 (1): 33–44.

Zhang, M. K., Liu, Z. Y. and Wang, H. (2010): Use of single extraction methods to predict bioavailability of heavy metals in polluted soils to rice. *Communications in Soil Science and Plant Analysis*, vol. 41 (7): 820–831.

# 8

# Bio - IPM Techniques: The Only Way to Produce Toxins Free Food

*S. S. Dhaka*

*Krishi Vigyan Kendra, Pilibhit, Sardar Vallabh Bhai Patel University of Agriculture & Technology, Modipurum, Meerut (U.P.), India*

## Integrated Pest Management: An Cardinal System of Plant Protection

Agricultural production in India increased at a high rate during the last five decades, leading to an era of food self-sufficiency. The remarkable growth was achieved by adopting new technologies in the form of high yielding crop varieties, chemical fertilizers and pesticides as well as from the expansion of cropped area. Nevertheless, the growth in agricultural production needs to be sustained to meet the food demand of ever increasing population. Since the scopes for bringing additional land under cultivation are limited, growth in agricultural production is possible only by increasing productivity. In other words, technology will be a key to future growth of agriculture.

There are many abiotic and biotic stresses which are responsible for the loss in productivity of the crops. As the abiotic factors are natural and environmental so they are beyond our control but the biotic stresses may be controlled easily. Amongst the biotic stresses, insect, pests, diseases and weeds cause considerable damage to potential agricultural production. Evidences indicate that pests cause 25% loss in rice, 5-10% in wheat, 30% in pulses, 35% in oilseeds, 20% in sugarcane and 50% in cotton. The losses though cannot be eliminated altogether, these can be minimized. Until recently, chemical pesticides were increasingly relied upon to limit the production losses. Pesticide use in India increased from a mere 15 g/ha of gross cropped area in 1955-56 to 90 g/ha in 1965-66. Introduction of green revolution technologies in mid-1960s gave a fillip to pesticide use, and in 1975-76, it had increased to 266 g/ha, and reached a peak of 404 g/ha in 1990-91. Although, there is a paucity of reliable time-series information on pest-induced production losses, anecdotal evidences suggest that there is still increase in losses despite increase in the pesticide use. The paradox is explained in terms of rising pest problem, technological failure of chemical pesticides and changes in production systems. Nevertheless, pesticide use has started declining since 1990-91, reaching 265g/ha in 1998-99, without much affecting the agricultural productivity. The Government of India had adopted Integrated Pest Management (IPM) system as a cardinal principle of plant protection in 1985. But adoption of IPM has not been found encouraging due to lack of availability of sufficient quantity of desired bio-pesticides. At present bio-pesticides capture hardly 2 % of the agrochemical market.

## Probable Cause of Less Market Capturing by Bio-Pesticides

Most of the farmers are dependent on the pesticide sellers for the protection of their crops against pests and diseases. When they face any problem they take their crop samples to these retailers and want a solution from them. These retailers remain under the influence of various private limited companies who give them more margins for selling their products. So they give farmers the products which fetch them more profits rather than going for the actual need of the farmers. Many times they give many additional pesticides to the farmers which are of no use for them.

## Harmful Impacts of Chemical Pesticides on Humans as well as Environment

Some chemical pesticides pose a potential risk to humans and other life forms resulting unwanted side effects to the environment. No segment of the population is completely protected against exposure to highly toxic pesticides and from their potentially serious health effects. A disproportionate burden of pesticide load is shouldered by the people of developing countries. The high risk groups exposed to pesticides are production workers, formulators, sprayers, mixers, loaders and agricultural farm workers. During manufacture and formulation, the possibility of hazards may be higher due to lack of awareness of poor as well as illiterate workers about the risk of toxic chemicals. Chemical pesticides pollute the tissues of virtually every life form on the earth, the air, the lakes and the oceans, the fishes that live in them and the birds that feed on the fishes. Certain pesticides are termed as endocrine disruptors which elicit their adverse effects by mimicking or antagonizing natural hormones in the body. It has been postulated that their long-term, low-dose exposure is creating more harm to human health such as immune suppression, hormone disruption, diminished intelligence, reproductive abnormalities and cancer. These chemical pesticides cause harmful effects on human health due to contamination of food commodities, soils, waters, turfs, and other vegetations. In addition to killing targeted insects or weeds, chemical pesticides are also causing problems to other beneficial organisms like birds, fishes, beneficial insects and non-target plants. Contamination of both surface and ground water contamination is another big problem throughout the world since the chemical pesticides reach in to water through either runoff or leaching from soils as well as treated plants. Heavy treatment of soil with pesticides reduces the populations of beneficial soil microorganisms. Soil scientist Dr. Elaine Ingham rightly pointed out that the loss of both beneficial bacteria and fungi degraded the soils. Overuse of chemical fertilizers and pesticides have effects on the soil organisms that are similar to human overuse of antibiotics. Indiscriminate use of chemicals might work for a few years, but after a while, there will not be enough beneficial soil organisms to make available the nutrients. Plants depend on a variety of soil microorganisms to transform atmospheric nitrogen into nitrates which they can use. Certain herbicides disrupt this process such as triclopyr inhibits soil bacteria that transform ammonia into nitrite; glyphosate reduces the growth and activity of free-living nitrogen-fixing bacteria in soil and 2, 4-D reduces nitrogen fixation by *Rhizobium* bacteria that live on the roots of bean plants, reduces the growth and activity of nitrogen-fixing blue-green algae, and inhibits the transformation of ammonia into nitrates by soil bacteria. Similarly, Mycorrhizal fungi grow with the roots of many plants and aid in nutrient uptake. These fungi can also be damaged by applying excessive toxic herbicides in the soil. Pesticide sprays can either directly hit non-target vegetation, or can drift or volatilize from the treated area and contaminate air, soil, and non-target plants.

## Solution to Avoid Harmful Impacts of Chemical Pesticides

The solution to above mentioned pesticide adversities lies in the implementation of Bio Integrated Pest Management (IPM) system, which combines the use of different pest control strategies such as adoption of proper cultural practices, cultivation of resistant varieties, and following of more biological and less chemical control techniques. IPM is no doubt more complex in nature for the producer to follow and implement, since it requires skills in pest monitoring and understanding of the pest dynamics.

Research has generated new technologies using naturally occurring enemies of insect pests (parasitoids, predators and pathogens) for use in IPM. Some important commercially available products include *Trichogramma, Bracons, Crysoperla carnea, Crytaemus montrouzieri, Bacillus thuringiensis, Bacillus sphaericus,* Nuclear polyhedrosis viruses (NPV) and *Trichoderma*. In addition, a number of plant products such as *Azadirachtin, pyrethrum*, nicotine (neem) etc. are also available as biopesticides. In India, more than 160 natural enemies have been studied for their utilization against insect pests. Technologies have been standardized for multiplication of 26 egg parasitoids, 39 larval/nymphal parasitoids, 26 predators and 7 species of weed. The Directorate of Plant Protection and Quarantine, Ministry of Agriculture, Government of India, has evolved location-specific IPM packages for both the *Kharif* and *Rabi* crops in consultation with IPM experts from the Indian Council of Agricultural Research, State Agricultural Universities and the State Departments of Agriculture.

## Useful Parameters for Successful IPM

Studies on following parameters are very essential for obtaining success in IPM.

### i) Technical Efficacy

Technical efficacy of IPM is judged on two criteria: change in the pesticide use and yield change over the conventional chemical control technique. As far as change in pesticide use is concerned, it is the basic goal of IPM to reduce the conventional chemical pesticide use by using efficient pesticide which is well established under experimental as well as field conditions. Its effect on yield could be either way. Results presented below (Table -1) suggest substantial yield saving advantage of IPM over chemical control in food as well as non-food crops:-

**Table 1: Cost of crop protection and the respective saving using IPM techniques**

| Crop | Cost of Plant protection (Rs/ha) | | Saving (Rs/ha) |
|---|---|---|---|
| | Conventional | IPM | |
| Cotton | 12600 | 2520 | 10080 |
| Chillies | 37600 | 5000 | 32600 |
| Pigeon pea | 3760 | 800 | 2960 |
| Groundnut | 3760 | 800 | 2960 |
| Castor | 5000 | 1000 | 4000 |
| Paddy | 5000 | 600 | 4400 |

*(Annual reports of IPM 2005-06)*

### ii) Economic Feasibility

Technical feasibility is necessary but not a sufficient condition for commercialization and adoption of a technology. The necessary condition is the net benefits it entails to the producers over the conventional technology. Net benefits can be measured in terms of the difference in per hectare net revenue due to application of new technology and/or changes in unit cost of production. Studies suggest IPM as a cost-effective technology. The magnitude of net benefits, however, would depend on the type of input used in IPM package, its application rate and price. Evidences show that even under experimental conditions some technically feasible IPM packages turn out to be economically infeasible because of higher prices of some of its constituents. The inference is 'IPM has the potential to substitute chemical pesticides without demanding any additional resources and without having any adverse effects on agricultural productivity. Nevertheless, inputs prices are an important determinant of the economic feasibility of IPM, and any increase in prices of critical inputs may upset its economics'.

### iii) Socio-economic and Policy Issues

Despite its techno-economic superiority over conventional chemical control, adoption of IPM remains restricted to hardly 2 percent of the area treated with plant protection inputs. This estimate is based on the informed opinions of the researchers, extension personnel and policy makers. The structure of agrochemical market also suggests a similar level of adoption; biopesticides share only 2 percent of the agrochemical market in India. There could be a number of technological, social, economic, institutional and policy factors restricting large scale adoption of IPM.

### iv) Technology Characteristics are Important Determinants of Adoption

The characteristics of technology have an important role in farmers' adoption decisions. IPM draws heavily on complementarities and interactions of different methods of pest control (chemical, biological, cultural and mechanical), and each of the components has its own specific characteristics and requirements for application. This makes IPM a complex technology. Generally, the farmers adopt those components that show immediate effect, and are easily available. Biopesticides comprise a major component of IPM. Most of the biopesticides are host-specific, slow in action and have short shelf-life. Besides, application of some of the components is labour intensive compared to conventional chemical control. In other words, farmers are risk averse and such technological characteristics create an apprehension among the farming about their efficacy to control pests. The complexity of IPM necessitates active involvement of stakeholders (researchers, extension workers and farmers) to alleviate apprehensions through participatory/ adaptive research trials. The major issues that the researchers would be confronting in the decades to come include basic research for development of broad-spectrum biological pesticides and improvements in their efficacy and shelf-life. At present, problems of insecticide resistance, resurgence and secondary pest outbreak are not reported against biological substitutes. Maintenance of this property would require sustained research efforts. Biopesticides based on predators, parasites, viruses, fungi, etc. are sensitive to chemical pesticides. This warrants research emphasis on development of bio-pesticides having better compatibility with chemical pesticides. Genetic engineering for resistance breeding will remain a gray area for long. Biotechnology has got tremendous potentialities for developing biopesticides.

### v) Community Participation is Key to Success of IPM

Pest has the characteristics of a detrimental common property resource. It does not recognize spatial boundaries. In other words, successful pest control demands collective efforts. Yet, most of the times pest control efforts are individualistic, giving rise to a number of pest control related problems, such as pest resistance, resurgence and secondary outbreak, destruction of natural enemies of insect pests and other beneficial insects. Collective pest management assumes greater significance in the context of IPM. There are a number of management practices such as observance of synchronicity in sowing dates, use of resistant varieties, crop rotations, etc. that require close cooperation among farmers to achieve maximum pest control efficiency. Further, IPM relies on inputs derived from living organisms, and the application of different control methods in a locality, in particular chemical pesticides, would adversely affect the activities of the biological inputs. Though, a majority of the farmers could be aware of the benefits of collective action, a number of socio-economic factors act as a disincentive to participate in it.

Community participation is key to successful adoption of IPM, and needs to be sustained by devising an appropriate exit policy. Local bodies, such as Panchayats, Non-Governmental Organizations (NGO), Self Help Groups (SHG), etc. should be encouraged to shoulder this responsibility. Incentives and awards should be given to those farmers/groups who are following IPM approach.

### vi) Supply of Biopesticides is Critical to Sustainability of IPM

As noticed earlier, biopesticides capture only 2% of the agrochemical market, although the mass production standards and techniques have been developed for a number of biopesticides. Further, most of the production takes place in public sector units. So is their distribution. Of over 400 biocontrol laboratories in India, 70% are in the domain of public sector. Most of the laboratories are small and cater to the location-specific needs only of a small area. The average gross cropped area per biocontrol laboratory is large. This shows that production of biopesticides is thinly spread. Nevertheless, the continental dimensions of the agricultural sector offer vast scope for expansion of biopesticide industry. Some inherent technical characteristics of bio-pesticides however act as disincentive to the entry of private sector. Unlike chemical pesticides, most of the bio-pesticides are not broad spectrum and are slow in action. Many of these like *Trichogramma* and *Crysoperla* have a short shelf-life, ranging from a few weeks to few months. Thus, production of biopesticides is fraught with risk. Other constraints in expansion of biopesticide industry are uncertain demand, and lack of appropriate infrastructure for transportation, storage and marketing. Rural unemployed and educated youths should be encouraged to establish small-scale biopesticide production units at village or block level. Measures, such as training to the potential entrepreneurs, provision of institutional credit, subsidies, insurance against low off take of inputs due to low pest infestation, and exemption from taxes and duties would stimulate production of biopesticides. Further, bio-pesticide manufacturing units are under strict registration and quality control requirements. The process of registration is cumbersome and costly, which discourages potential entrepreneurs.

Considering the role of biopesticides in ecological conservation and human health safety, registration requirements should be relaxed, though without reconciliation with quality standards.

### vii) Enforcement of Pesticide Regulations will Help Improve Adoption of IPM

In recent years, the central government has banned a number of pesticides for use in agriculture in consideration of their adverse effects on environment and human health. Despite this, many of these are available in the market. For example, DDT and BHC, which are permitted for the use for malaria control, are widely used in agriculture. Further, many pesticides that have been banned elsewhere in the world are available to Indian farmers. Lower prices of such pesticides induce farmers to use them. A number of spurious pesticides are available in the market because of lack of strict enforcement of regulations and/or regulatory loopholes. Strict enforcement of the regulations governing production, use, distribution and quality of pesticides would help weed out spurious elements from the industry and would benefit the farmers.

### viii) Economic Incentives will Encourage Farmers Switching Over to IPM

Price of a technology is an important determinant in farmers' decision for its adoption. At present, bulk of the supply of biopesticides comes from public sector, often provided at subsidized prices under IPM programmes. The evidences show that benefits of adoption of IPM are marginally higher than the conventional chemical pest control. An increase in the price of biopesticides due to cost considerations or withdrawal of subsidies would upset the economics of IPM. Since biopesticides generate considerable social and environmental benefits, the government should think of classifying them into 'green box' for provision of subsidies. Simultaneously, the incentives/subsidies, if any, on synthetic pesticides be withdrawn and the resultant savings be diverted towards promotion of IPM. Linking of agricultural credit and insurance to IPM can also facilitate its faster diffusion. Another alternative is to make production and use of chemical pesticides unattractive through fiscal instruments of taxes, excise duties, sales taxes, etc. on intermediary inputs and final output. The decline in pesticide use during the early 1990s was on account of imposition of heavy taxes on pesticide industry. The pesticide industry, which has established strong market over the last three decades, may resist it, but it may be pursued to switch over to production of safer pesticides and biopesticides. Withdrawal of subsidies on chemical pesticides and diversion of the same towards production and use of biopesticides, and linking institutional credit and insurance with IPM adoption would induce farmers switching over to IPM.

### ix) Development of Market for Pesticide-Free Products is Necessary

Economic incentives may not be sustained for a long. An alternative is to develop markets for pesticide-free or low-pesticide residue produce by creating consumer awareness about health benefits of such produce. At present, there are no premium markets and standards for organic food in India. Since in the short-run there is a possibility of shortfall in yield on switching over to IPM, farmers even if they are willing to adopt IPM may not do so. In developed countries market for pesticide free products is developing and these products fetch premium prices. This however is lacking in India. This would require not only development of certification procedures and labeling system to gain confidence of the consumers. The cost of certification is high for an individual farmer. The cost can be brought down considerably if a group approach is followed. Evolving simple and cost-effective certification and labeling systems to enable farmers to produce pesticide-free products and to gain confidence of the consumers will boost adoption of IPM.

Integrated Pest Management (IPM) is an ecologically based strategy that focuses on long-term solution of the pests through a combination of techniques such as biological control, habitat manipulation, modification of agronomic practices, and use of resistant varieties. Embracing a single tactic to control a specific organism does not constitute IPM, even if the tactic is an essential element of the IPM system. Integration of multiple pest suppression techniques has the highest probability of sustaining longterm crop protection. Pesticides may be used to remove/prevent the target organism, but only when assessment with the help of monitoring and scouting indicates that they are needed to prevent economic damage. Pest control tactics, including pesticides, are carefully selected and applied to minimize risks to the human health, beneficial and non-target organisms, and environment in the context of crop protection, sustainability refers to the substitution of chemicals and capital with farm grown biological inputs and knowledge, aimed at reduction in the cost of production without lowering the yields. Sustainability builds on the current agricultural achievements, adopting a sophisticated approach that can maintain high yields and farm profits without degrading the resources. Sustainable agriculture is a reality based on the human goals and on the understanding of the long-term impact of human activities on the environment and on other species. This philosophy combines the application of prior experience and the latest scientific advancements to create integrated, resource-conserving, equitable farming systems. The systems approach minimizes environmental degradation, sustains agricultural productivity, promotes economic viability in both the short and long run, and maintains quality of the life. Sustainable farming practices commonly include:

Crop rotations that mitigate weeds, disease, insect and other pest problems; provide alternative sources of soil nitrogen; reduce soil erosion; and reduce risk of water contamination by agricultural chemicals · Pest control strategies include integrated pest management techniques that reduce the need for pesticides by practices such as scouting/ monitoring, use of resistant cultivars, timing of planting, and biological pest controls, increased mechanical / biological weed control; more soil and water conservation practices; and strategic use of green manures and use of natural or synthetic inputs in a way that poses no significant hazard to humans or the environment.

## Tools of IPM

### Monitoring

Crop monitoring that keeps track of the pests and their potential damage, is the foundation of IPM. This provides knowledge about the current pests and crop situation and is helpful in selecting the best possible combinations of the pest management methods. Pheromone traps have got advantage over other monitoring tools such as light and sticky traps (Fig.1). Being selective to specific pest, they have proven their usefulness in large scale IPM validations in cotton, basmati rice, chickpea and pigeon pea (Fig.2).

**Fig.1:** Pheromone traps for monitoring insects

**Fig.1:** Pheromone traps for monitoring insects

**Fig.3**: Mass trapping of adult males of many insects is done with the help of pheromone traps. Many of the insets get killed in these traps without the use of any chemical pesticides.

**Fig.4:** Use of traps and mating disruptors for the control of stem borer in Rice ecosystem

## Pest resistant Varieties

Breeding for pest resistance is a continuous process. At the same time the pests also, particularly the plant pathogens, co-evolve with their hosts. Thus, gene transfer technology is useful in developing cultivars resistant to insects, plant pathogens and herbicides. An example of this is the incorporation of genetic material from *Bacillus thuringiensis* (Bt), a naturally occurring bacterium, in cotton, corn, and potatoes, which makes the plant tissues toxic to the insect pests. Scientific community is impressed by its huge potential in managing the pests, but is also concerned about the possibility of increased selection pressure for resistance against it and its effects on non-target natural fauna. However, due to ethical, scientific and social considerations, this potential technology has been surrounded by controversies.

### Cultural Pest Control

It includes crop production practices that make crop environment less susceptible to pests. Crop rotation, fallowing, manipulation of planting and harvesting dates, manipulation of plant and row spacing, and destruction of old crop debris are few examples of cultural methods that are used to manage the pests. Planting of cover crops, nectar producing plants and inter-planting of different crops to provide habitat diversity to beneficial insects are important management techniques. Cover crops, often legume or grass species, prevent soil erosion and suppress weeds. A cover crop can also be used as a green manure, which is incorporated in the soil to provide nitrogen and organic matter to the subsequent crop. When incorporated in the soil, some cover crops of the *Brassica* family have the ability to suppress nematode pests and wilt diseases. Left in the field as residues, rye and wheat provide more than 90% weed suppression. Cultural controls are selected based on knowledge of pest biology and development.

**Fig. 5:** Summer deep ploughing to kill overwintering pest's stages

**Fig. 6:** Mustard as trap crop in cabbage to manage different pests. The pests attacking the cabbage come to the mustard crop first and they may be controlled easily and the main crop escapes the damage caused by pests.

## Physical or Mechanical Control

These are based on the knowledge of pest behaviour. Placing plastic-lined trenches in potato fields to trap migrating Colorado potato beetles is one of the examples of the physical control. Shaking of the pigeon pea plant to remove *Helicoverpa* larvae is a common practice in pigeon pea growing areas. Hand picking of insect pests is perhaps the simplest pest control method. Installation of dead as well as live bird perches in cotton and chickpea fields has proved effective in checking the bollworm infestation. Using mulches to smother weeds and providing row covers to protect plants from insects are other examples.

**Fig. 7:** Light trap to control nocturnal insects.

**Fig. 8:** Adhesive or yellow trap to for aphids.

## Biological Control

These include augmentation and conservation of natural enemies of pests such as insect predators, parasitoids, parasitic nematodes, fungi and bacteria. In IPM programs, native natural enemy populations are conserved, and non-native agents may be released with utmost caution. *Trichogramma* sp. is the most popular parasitoids being applied on a number of host crops. A number of microorganisms such as *Metarrhizium*, *Beauveria*, Nuclear Polyhedrosis Virus (NPV), *Trichoderma* sp., *Verticillium* sp., *Aspergillus* sp., *Bacillus* sp. and *Pseudomonas* sp. that attack and suppress the insect pests and plant pathogens have been exploited as biological control agents.

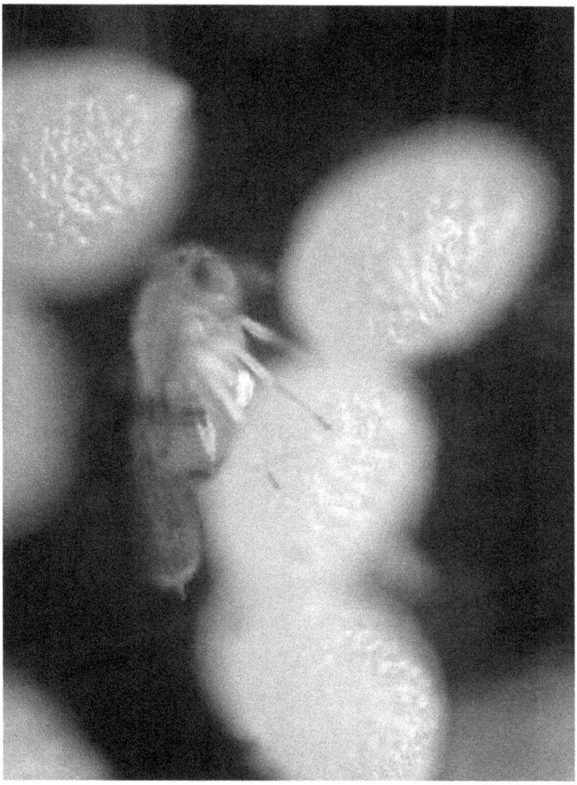

**Fig. 9:** Trichogramma, an egg Parasitoid of pests

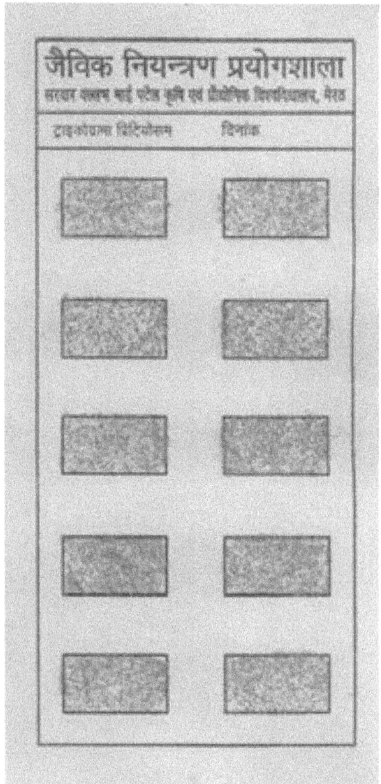


**Fig.10:** Tricho card to release Trichogramma in field

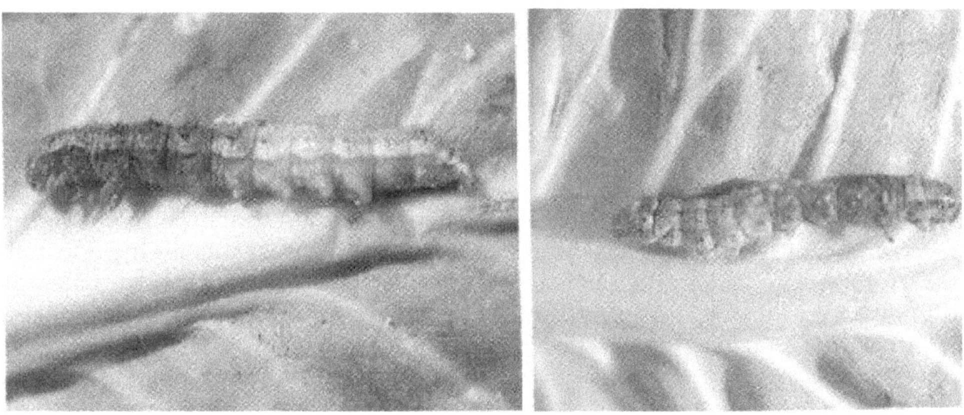

**Fig.11:** Bacillus thuringiensis (a bacterial pathogen) infected Helicoverpa larva

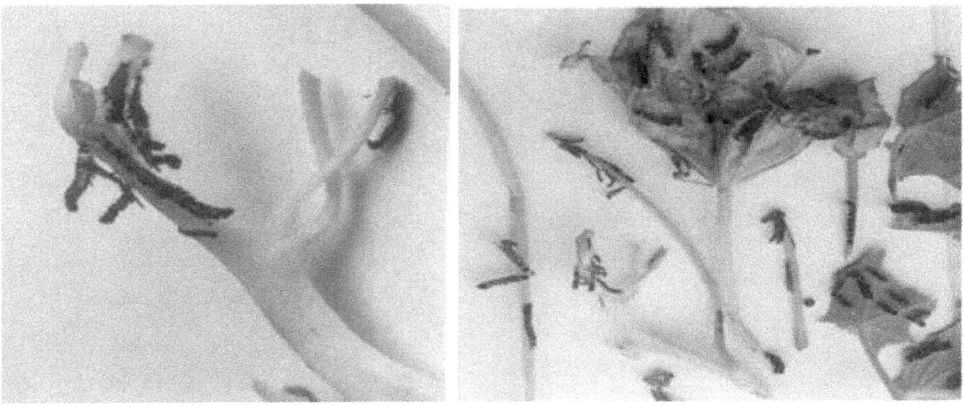

**Fig.12:** Nuclear Polyhedrosis Virus (NPV) - a viral pathogen infected *Helicoverpa* larva

*Metarrhizium* infected white grub

**Fig.13**

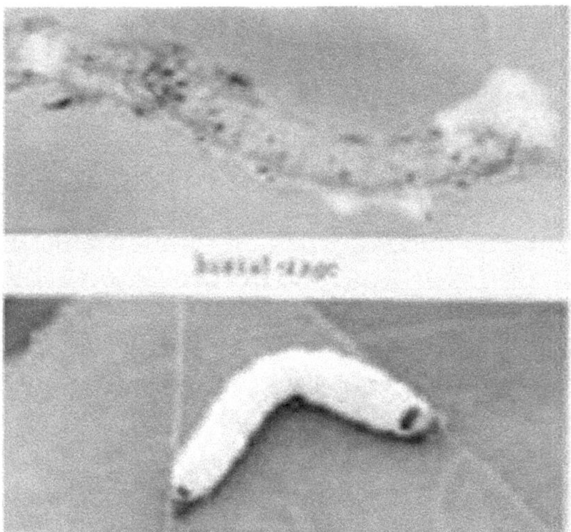

*Beauveria* infected *Helicoverpa*

**Fig.14**

## Chemical control

Pesticides are used to keep the pest population below economically damaging level when the pests cannot be controlled by other means. Pesticides include both the synthetic pesticides and plant-derived pesticides. Synthetic pesticides include a wide range of man-made chemicals. These are easy to use, fast-acting and relatively inexpensive. Ideally, pesticides should be used as a last resort in IPM programs because of their potential negative effect on the environment. Pesticides with the least negative impacts on non-target organisms and the environment are most useful. Fortunately, new generation pesticides with novel modes of action and low environmental effects are being developed and registered for the use. Pesticides that are short-lived or act on one or a few specific organisms fall in this class. Economic threshold assessment is based on the concept that most plants can tolerate at least some pest damage. Much research has been done to determine the damage thresholds for a variety of crops and pest situations, yet the studies are inconclusive. In an IPM program where the economic threshold is known, chemical controls are applied only when the pest's damaging capacity is nearing to the threshold, despite the application of other alternative management practices. Botanical pesticides can be prepared in various ways. They can be as simple as raw crushed plant leaves, extracts of plant parts, and chemicals purified from the plants. Pyrethrum, neem, tobacco, garlic and pongamia formulations are some of the examples of botanicals. Some botanicals are broad spectrum pesticides. Botanicals are generally less harmful to the environment, because of their quick degrading property. They are less hazardous to transport. The major advantage is that these can be formulated on-farm by the farmers themselves.

## Economic Viability of IPM

IPM practices are economical viable for the farmers. Here an example is given in which a research trial was conducted to test various IPM modules against fruit and shoot borer in okra crop. The cost benefit ratio of different modules was also worked out (Table-1 & 2).

**Table-1: Treatment combinations in different modules**

| S. No. | Modules | Treatments combinations |
|---|---|---|
| 1 | $T_1$ ($M_1$) | **Bio-intensive** |
| | | Application of neem cake @ 250 kg/ ha at the time of field preparation. |
| | | Seed treatment with *Trichoderma harzianum* @ 10 gm /kg seed. |
| | | Installation of pheromone traps @ 5trap/ acre for monitoring the population of shoot and fruit borers, at 15 DAS. |
| | | Four realeases of egg parasite *Trichogramma chilonis* @ 1 lac /ha. for shoot and fruit borer starting from pre flowering and subsequently at weekly interval . |
| | | Three spray of NSKE @ 5% against jassid, white fly and aphids at 15, 30 and 45 DAS. |
| | | One spray of *Beuveria bassiana* @ 2.5 kg/ha. at 60 DAS. |
| 2. | $T_2$ ($M_2$) | **Mechanical and Bio-intensive + Cultural** |
| | | All six practices of $M_1$ |
| | | 7. Installation of delta and yellow sticking traps @ 10 trap / acre for monitoring the population of sucking insect pests viz. white fly etc. After one week of sowing. |
| | | 8. Collection and destruction of egg masses, grubs and damaged shoot and fruits. |
| | | 9. Roughs out of YVMV and other diseased plants at 15 days interval. |

| | | |
|---|---|---|
| 3. | $T_3$ ($M_3$) | **Bio-intensive + Mechanical + Cultural + Chemical**<br>All nine practices of $M_2$<br>10. One spray of Imidacloprid 17.8 SL @ 0.5 ml/lit. at 70 DAS. |
| 4. | $T_4$ ($M_4$) | **Farmers practices prevalent in the area **** |
| 5. | $T_5$ ($M_5$) | **Untreated control** |

** In this modules farmers' own way of managing the pests, which comprised of only chemicals were followed. One spray of abate (temphos 50%, mosquito larvicide) @ 0.04% at 25 DAS; one spray of monocrotophos (monitor) 40 EC @ 0.05% at 45 DAS; one spray if acephate 75 SP @ 0.07% at 60 DAS; one spray of megha (chlorpyriphos 50% + Cypermethrin 5%) @ 0.04% at 75 DAS; one spray of cypermethrin 25 EC (mortal) @ 0.004% at 90 DAS; one spray of acetamiprid 20 SP (pride+alphamethrin+phosphix) @3 ml / litre at 105 DAS; one spray of megha (chlorpyriphos 50% + Cypermethrin 5%) + Cypermethrin 25 EC (mortal) + acetamiprid 20 SP (pride) @ 3 ml /litre at 120 DAS.

**Table 2: Cost benefit ratio of different treatments for the management of major insect pests of okra during *Kharif*, 2008.**

| Modules | Marketable yield (q/ha) | Yield gain over control (q/ha) | *Value of yield gain (Rs./ ha) | **Cost of treatment (Rs./ ha) | Net profit / ha (Rs. / ha) | Cost benefit ratio |
|---|---|---|---|---|---|---|
| **Module-I** | 41.55 | 12 | 14400 | 4188 | 10212 | 1 : 2.44 |
| **Module-II** | 48.29 | 18.74 | 22488 | 5538 | 16950 | 1 : 3.06 |
| **Module-III** | 72.52 | 42.97 | 51564 | 5668 | 45896 | 1 : 8.10 |
| **Module-IV** | 60.43 | 30.88 | 37056 | 5880 | 31176 | 1 : 5.30 |
| **Module-V** | 29.55 | 0 | 0 | 0 | 0 | - |

It is quite evident from the above example that the highest cost benefit ratio of 1:8.10 was obtained when all the practices of IPM were followed properly while a lower cost benefit of 1:5.30 was found with the farmers' practice wherein several chemical insecticides were used in higher quantity. It is therefore proved that IPM is economical viable and can give farmers a better economic return if used properly.

## Strategies for IPM Implementation

The IPM packages tested at several research centers vis-a-vis the farmers' practices indicate superiority of the former. IPM practices enabled reduction in the number of chemical sprays. IPM system also resulted in increase of natural enemies by three-fold, reduced the insecticide and environmental pollution. An integrated strategy for the management of major pests and diseases is possible by

(i) breeding new varieties with built-in resistance, (ii) evolving efficient methods of pest control through pest surveys and monitoring, and (iii) biological control of pests with the help of conservation and augmentation 20 of natural enemies like parasites, predators and insect pathogens. Economically viable integrated pest management strategies have been developed for the control of major pests in rice, cotton, pulses, sugarcane, etc. Control of *Pyrilla* and top borer of sugarcane, mealy bug of coffee, lepidopterous pests affecting cotton, tobacco, coconut, sugarcane, etc. are few examples where success has been achieved through the release of biocontrol agents. A major achievement has been the development of mass rearing technology for biocontrol agents such as *Trichogramma* sp., *Chrysoperla* sp. and nuclear polyhedrosis viruses (NPV) of *Helicoverpa* and *Spodoptera*.

## Major Obstacles

Although, IPM has been accepted as the most attractive option for the protection of crops from the ravages of pests, implementation at the farmer's level has been limited. Pesticides continue to dominate and their injudicious use represents the greatest threat to IPM. For an effective implementation strategy, it is necessary to identify the obstacles to its dissemination, some of which are:

- Low awareness and innovativeness of extension personnel and target groups
- Inadequate interaction between research and extension agencies
- Problem of timely and adequate supply of quality inputs, including biocontrol agents and Bio-pesticides
- Complexity of IPM vs simplicity of chemical pesticides
- The dominant influence of pesticide industry
- Non-availability of location-specific IPM modules for many crops

## Essentials for implementation

- Availability of location-specific IPM modules, which are ecologicallysound, economically viable and socially acceptable
- High level of target group participation
- Area-wide dissemination strategy
- Removal of obstacles in dissemination of IPM
- Measuring, evaluating and publicizing the impacts of IPM.

Conservation of natural enemies of pests and their augmentation is of prime importance. Besides, the intrinsic property of renewability, reversibility and resilience of botanicals and biopesticides make them most dependable tools for sustainable IPM. Hence, to maintain ecological balance and to manage the pests, the use of bio-agents and biopesticides/botanicals must receive priority attention.

## Conclusion

There is an emerging consensus that modern petrochemical-based farming is unsustainable and there is a need to develop and promote ecological approaches to food production. The most obvious and apparently environment-friendly alternative to pesticides is to follow the naturally occurring biological approaches. Many plant species have been reported to possess pesticidal and pest growth inhibiting properties, but their potential remains untapped by the industry. Holistic planning provides farmers with the management tools they need to manage biological complex farming systems in a profitable manner. A successful IPM program requires time, money, patience, short - and long term planning, flexibility and commitment. The research managers must spend time on self-education and making contacts with extension and research personnel to discuss farming operations, which vary widely. This would aid in developing integrated plans. The government could create policy environment for promotion of IPM. The central and state governments must take lead in changing the pest control picture through measures that would make chemical control less attractive through legislation, regulatory and fiscal measures. The Indian Council of Agricultural Research (ICAR) and the Department of Agricultural Research and Education of the Ministry of Agriculture, Government of India, are committed to the development and promotion of IPM in the country. It is the top priority of the ICAR and the Government of India to provide safe and effective technologies to protect against unacceptable losses due to insect pests, weeds and diseases.

India has successfully reduced pesticide consumption without adversely affecting the agricultural productivity. This was facilitated by appropriate policies that discouraged pesticide use, and favoured IPM application. Despite it, adoption of IPM is low owing to a number of socio-economic, institutional and policy constraints. On the supply side, lack of commercial availability of biopesticides and inappropriate institutional technology transfer mechanisms are the critical impediments to increased application of IPM. The presence of private sector in biopesticides production and marketing is marginal, and needs to be improved through economic incentives. On the demand side, farmers though are aware of technological failure of pesticides to control pests, and their negative externalities to environment and human health, pest risk is too high to experiment with newer approaches to pest management. IPM is a complex process and farmers lack understanding of biological processes of pests and their predators and methods of application of new technology components. The socio-economic environment of farming is also an important factor in adoption of IPM. There are a number of IPM practices that work best when applied by the entire community and in a synchronized mode. This is unlikely to happen without demonstrating benefits of group approach, and external motivation and support to the farmers. Though many technology programs are based on community approach, they do not have any proper exit policy to sustain the group approach. The IPM policy should also provide incentives to farmers to adopt IPM as a cardinal principle of plant protection.

## References


Anonymous (1997): Agricultural Statistics at a Glance. Directorate of Economics and Statistics, Ministry of Agriculture, Government of India, New Delhi.

Anonymous (2000): Agricultural Statistics at a Glance. Directorate of Economics and Statistics, Ministry of Agriculture, Government of India, New Delhi.

Anonymous (2001): Agricultural Research Databook. Indian Agricultural Statistics Research Institute, New Delhi.

Birthal, P.S. and D. Jha (1997): Socio-economic impact analysis of integrated pest management programmes. In: National Symposium on IPM in India – Constraints and Opportunities, Indian Agricultural Research Institute, New Delhi. October 23-24.

Carson, Rachel (1962): Silent Spring. Boston, Miffin Co.

Chakraborty, T. and R.B. Mallick (1995): Concept of sustainability in agriculture in the Indian context – Major issues and resource recycling. In: Proceedings of the National Symposium on Sustainable Agriculture in Sub-Humid Zone (March).

Charles, Francis and Garth, Youngberg (1990): Sustainable agriculture – An overview. In: Sustainable agriculture in temperate zones (Eds. CA Francis, CB Flora and LD King), New York, Wiley.

Dhaliwal, G.S. and Ramesh, Arora (1996): Principles of insect pest management. National Agricultural Technology Information Centre, Ludhiana.

Jayaraj, J. (1993): Biopesticides and integrated pest management for sustainable crop production.

In: Agrochemicals in sustainable agriculture (Ed. N.K. Roy), New Delhi, APC Publications.

Khader Khan, H. (1996): Integrated pest management and sustainable agriculture. Farmers and Parliament. 30 (2):15-17.

Krishnamurthy Rao, B.H. and Murthy, K.S.R.K. (1983):. Proceedings of National Seminar on Crop Losses due to Insect Pests. *Indian J. Entmol.* (Special issue), vols I-II, Hyderabad.

Mukhopadhyay, S.K. and R.B. Mallick (1995): Concepts and dimensions of sustainable agriculture. In: Proceedings of National Symposium on Sustainable Agriculture in Sub-Humid Zone, p.1-4.

Paroda, R.S. (1999): For a food secure future. Survey of Indian Agriculture. The Hindu.

Raheja, A.K. and Tewari, G.C. (1997): Awareness program on pesticides and sustainable agriculture. ICAR, New Delhi. p - 69

# Index

**Photo- 1: Flower decorated Shree Siddhivinayak Ganapati (p.1)**

**Photo-3: Flower wastes generated in the Farm (p. 2)**

**Photo- 2: Flowers sold by Flower Vendors (p. 2)**

**Photo- 4: Dumping of flower waste, on roadside (p.3 )**

**Photo - 5: Flower waste thrown in Ganga River (p.4)**

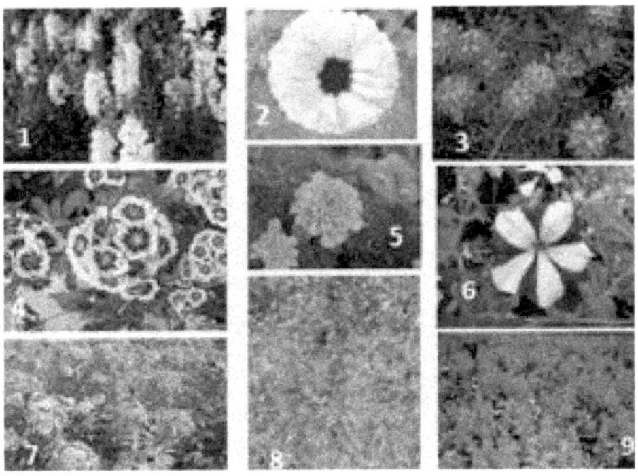

1. Antirrhinum, 2. Calendula, 3. Candytuft, 4. Dianthus, 5. Marigold, 6. Petunia, 7. Phlox, 8. Poppy, 9. Saliva Photo- 7: Seasonal flowers used for vermicomposting (p. 8)

Photo- 8: A. *magna millipede* (p. 14)

Photo-1: Bareilly Mandi. (p. 20 )

Photo-2: Paddy straw as packing waste in Bareilly M andi (p. 20)

Photo-4: Discussing with Mandi Incharge Mr. K. D. Sharma & Mr. Abdul Hussian (p. 22)

Photo-5: Waste on a mandi street (p. 22)

**Photo-6: Oranges as fruit waste in the Mandi (p. 23)**

**Photo-7: Vegetable waste of Mandi along with packing waste (p. 23)**

**Photo-8: Vermiculture and Vermicomposting Centre of the University (p. 24)**

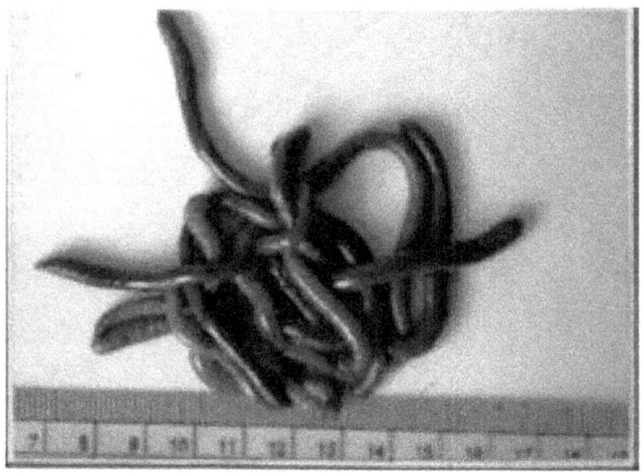

**Photo-9: Earthworms applied for vermicomposting (p. 24)**

***Photo-10: A bed of Vermiculture & Vermicomposting Centre showing Vermicomposting process (p. 25)***

**Photo-11: Vermicomposting bed showing proper level of moisture content (p. 25)**

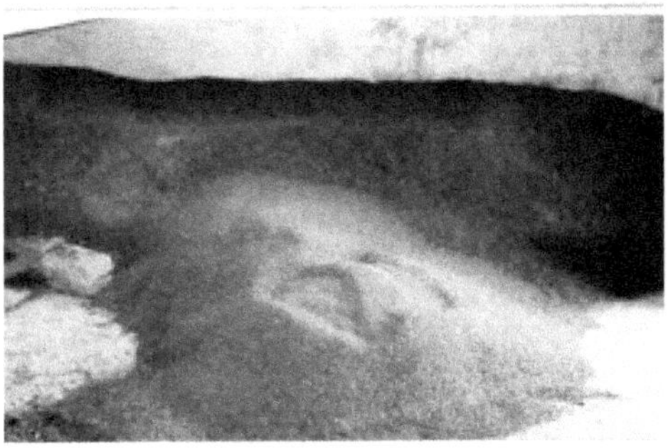

Photo-12: Ready Vermicompost from the vegetable and fruit waste of the Mandi (p. 27)

Fig. 1: Paper sheredder machine and basket of shredded paper (p. 34)

Fig. 2: Mother culture vermicompositing tank Bunch of earthworms in inset (p. 34)

Fig. 3: A series of experiments were conducted in plastic containers (p. 34)

Fig. 4: Free choice experiment (p. 34)

Fig. 8: Four Chambered tank used for paper waste vermicopositing (p. 34)

**Fig. 1 & 2: Generation and transportation of pressmud from sugar factory (p. 56)**

**Fig. 3: Curring of pressumed for vermimanure production (p. 57)**

**Pressmud in cement tank**

**Fig. 4: Vermimanure production in small scale (p. 59)**

**Pressumud Vermimanure**

**Fig. 4: Vermimanure production in small scale (p. 59)**

**Pressumud Vermimanure**

**Fig. 6: Vermimanur production in large scale (p. 59)**

**Fig. 7: Growth and yield characteristics of chilli (p. 64)**

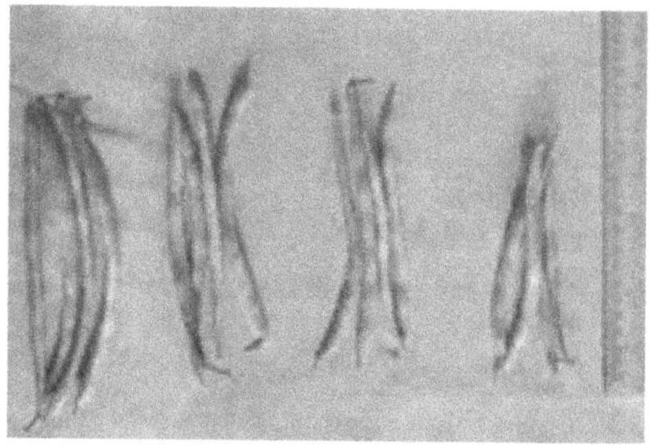

Fig. 8: Growth and yield characteristics of beans (p. 64)

FYM 10 $tha^{-1}$ +NPK 100%

Vermicompost 10 $tha^{-1}$ +NPK 100%)

Vermicompost 10 $tha^{-1}$ +NPK 100%)

Vermicomposted fly ash 10 $tha^{-1}$ +NPK 100%)

Fig. 5: Effect of vermicomposted fly ash on th grow1th of tomato crop (p. 75)

Fig.1: Pheromone traps for monitoring insects (p.99)

Fig.1: Pheromone traps for monitoring insects (p.99)

Fig.3: Mass trapping of adult males of many insects is done with the help of pheromone traps. Many of the insets get killed in these traps without the use of any chemical pesticides. (p. 99)

Fig.4: Use of traps and mating disruptors for the control of stem borer in Rice ecosystem (p.100)

Fig. 5: Summer deep ploughing to kill overwintering pest's stages (p. 101)

Fig. 6: Mustard as trap crop in cabbage to manage different pests. The pests attacking the cabbage come to the mustard crop first and they may be controlled easily and the main crop escapes the damage caused by pests(p.101)

Fig. 7: Light trap to control nocturnal insects. (p.102)

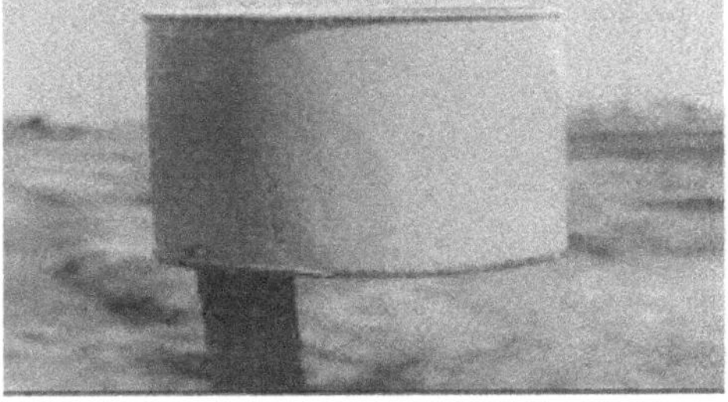

Fig. 8: Adhesive or yellow trap to for aphids. (p 102)

www.ingramcontent.com/pod-product-compliance
Ingram Content Group UK Ltd.
Pitfield, Milton Keynes, MK11 3LW, UK
UKHW021323190726
13845UKWH00003B/32

9 789390 371112